Suzuki GT 250 X7, GT 200 X5 & SB 200 Owners Workshop Manual

by Pete Shoemark

Models covered:

GT250 X7.	247cc.	May 1978 to August 1983
GT200 X5.	196cc.	March 1979 to January 1982
SB200.	196cc.	March 1979 to March 1983

ISBN 978 1 85010 068 3

(469-8S6)

Haynes Group Limited
Haynes North America, Inc

www.haynes.com

British Library Cataloguing in Publication Data

Shoemark, Pete
Suzuki GT250X7, GT200X5 & SB200 1978-1983 owners workshop manual.—(owners workshop manual)
1. Suzuki motorcycle
I. Title II. Churchill, Jeremy
629.28'775 TL448.S8
ISBN 1-85010-068-3

The manufacturer's authorised representative in the EU for product safety is:

HaynesPro BV
Stationsstraat 79 F, 3811MH Amersfoort, The Netherlands
gpsr@haynes.co.uk

Acknowledgements

Our thanks are due to Fran Ridewood and Co. of Wells, who supplied the machine featured in the photographs throughout this Manual. Special thanks are due to Mr A. Jerham and Mr P. Barber of Heron Suzuki (GB) Ltd, who supplied service information and advice on the X7, X5 and SB 200 models, and who checked the text of the Manual for technical accuracy.

Brian Horsfall assisted with the stripdown and rebuilding, and devised the ingenious methods for overcoming the lack of manufacturer's service tools. Les Brazier took the photographs which accompany the text; Mansur Darlington edited the text.

Finally, we would like to thank the Avon Rubber Company, who kindly supplied information and technical assistance on tyre fitting; NGK Spark Plugs (UK) Ltd for information on sparking plug maintenance and electrode conditions, and Renold Ltd for advice on chain care and renewal.

The machine shown on the front cover was loaned by Mr. R.D. Willis of Yeovil, to whom we would like to express our thanks.

About this manual

The purpose of this manual is to present the owner with a concise and graphic guide which will enable him to tackle any operation from basic routine maintenance to a major overhaul. It has been assumed that any work will be undertaken without the luxury of a well-equipped workshop and a range of manufacturer's service tools.

To this end, the machine featured in the manual was stripped and rebuilt in our own workshop, by a team comprising a mechanic, a photographer and the author. The resulting photographic sequence depicts events as they took place, the hands shown being those of the author and the mechanic.

The use of specialised, and expensive, service tools was avoided unless their use was considered to be essential due to risk of breakage or injury. There is usually some way of improvising a method of removing a stubborn component, provided that a suitable degree of care is exercised.

The author learnt his motorcycle mechanics over a number of years, faced with the same difficulties and using similar facilities to those encountered by most owners. It is hoped that this practical experience can be passed on through the pages of this manual.

Where possible, a well-used example of the machine is chosen for a workshop project, as this highlights any areas which might be particularly prone to giving rise to problems. In this way, any such difficulties are encountered and resolved before the text is written, and the techniques used to deal with them can be incorporated in the relevant Section. Armed with a working knowledge of the machine, the author undertakes a considerable amount of research in order that the maximum amount of data can be included in the manual.

Each Chapter is divided into numbered sections. Within these Sections are numbered paragraphs. Cross reference throughout the manual is quite straightforward and logical. When reference is made 'See Section 6.10' it means Section 6, paragraph 10 in the same Chapter. If another Chapter were intended, the reference would read, for example, 'See Chapter 2, Section 6.10'. All the photographs are captioned with a section/paragraph number to which they refer and are relevant to the Chapter text adjacent.

Figures (usually line illustrations) appear in a logical but numerical order, within a given Chapter. Fig. 1.1 therefore refers to the first figure in Chapter 1.

Left-hand and right-hand descriptions of the machines and their components refer to the left and right of a given machine when the rider is seated normally.

Motorcycle manufacturers continually make changes to specifications and recommendations, and these, when notified, are incorporated into our manuals at the earliest opportunity.

Contents

Right-hand view of Suzuki GT250 X7

Left-hand view of Suzuki GT200 X5

Introduction to the Suzuki SB200, GT200 X5 and GT250 X7

The Suzuki Motor Company Ltd have long been associated with the production of high performance two-stroke machines, the first of which appeared in the UK in 1963. The first of the twin cylinder models, the T10, was introduced in 1964. This model eventually evolved into the popular GT250 range during the mid-1970s.

By 1978, the general trend was the production of four-stroke machines. Many considered the two-stroke motorcycle to be too thirsty and to create an unreasonable amount of exhaust pollution for it to be a reasonable proposition for the coming years. In consequence, the new range of lightweight two-stroke twins was greeted with some surprise by the motorcycling world.

Each of the three models employs a variation of the same basic engine/gearbox unit, the SB200 having a piston-ported engine driving through a four-speed gearbox, the X5 using a more powerful version of the same engine and a five-speed gearbox, while the X7 uses a combined piston-port and reed valve induction system and is fitted with a six-speed gearbox to make the best use of its high power output.

The SB200 is fitted with drum brakes and is styled conservatively to provide a basic but well-equipped machine for the commuter, while the X5 and X7, with their cast alloy wheels, front disc brakes and more exciting styling, are designed to appeal to younger riders. Strict attention has been paid to keeping the weight as low as possible, in order to improve performance and handling. To enable it to compete with its closest rivals, the X5 is the only model imported into the UK that is fitted with electric start.

Following its usual policy of constant development, Suzuki introduced slightly modified versions of all three models. While the most significant alterations were for styling purposes, notably in the tank and side panel striping, there are some other differences which will require some care when ordering replacement parts. For example, the later versions of the X5 and X7 are fitted with 'dog-leg' handlebar levers and square handlebar master cylinder reservoirs. To assist the owner in identifying his machine the frame numbers with which each version began its production run are given below, with the model code, model name and approximate date of import into the UK. Note that it is not usually sufficient to identify a machine by giving its date of registration.

Model	Model code	Frame number	Approx. year
SB200	SB200	SB200-100001	Mar '79 – Jan '82
SB200	SB200X	SB200-103762	Jan '82 – Mar '83
X5	GT200EN	GT200-500001	Mar '79 – Jan '81
X5	GT200EX	GT200-509890	Jan '81 – Jan '82
X7	GT250EN	GT2502-500001	May '78 – Oct '80
X7	GT250EX	GT2502-519409	Oct '80 – Aug '83

Dimensions and weight

	SB200	GT200 X5	GT250 X7
Overall length	1980 mm (78.0 in)	1965 mm (77.4 in)	2005 mm (78.9 in)
Overall width	820 mm (32.3 in)	695 mm (27.4 in)	740 mm (29.1 in)
Overall height	1070 mm (42.1 in)	1030 mm (40.6 in)	1065 mm (41.9 in)
Wheelbase	1295 mm (51.0 in)	1300 mm (51.2 in)	1310 mm (51.6 in)
Ground clearance	175 mm (6.9 in)	175 mm (6.9 in)	160 mm (6.3 in)
Dry weight	116 kg (256 lb)	123 kg (271 lb)	128 kg (282 lb)
Seat height	790 mm (31.1 in)	790 mm (31.1 in)	—

Safety first!

Professional motor mechanics are trained in safe working procedures. However enthusiastic you may be about getting on with the job in hand, do take the time to ensure that your safety is not put at risk. A moment's lack of attention can result in an accident, as can failure to observe certain elementary precautions.

There will always be new ways of having accidents, and the following points do not pretend to be a comprehensive list of all dangers; they are intended rather to make you aware of the risks and to encourage a safety-conscious approach to all work you carry out on your vehicle.

Essential DOs and DON'Ts

DON'T start the engine without first ascertaining that the transmission is in neutral.

DON'T suddenly remove the filler cap from a hot cooling system – cover it with a cloth and release the pressure gradually first, or you may get scalded by escaping coolant.

DON'T attempt to drain oil until you are sure it has cooled sufficiently to avoid scalding you.

DON'T grasp any part of the engine, exhaust or silencer without first ascertaining that it is sufficiently cool to avoid burning you.

DON'T allow brake fluid or antifreeze to contact the machine's paintwork or plastic components.

DON'T syphon toxic liquids such as fuel, brake fluid or antifreeze by mouth, or allow them to remain on your skin.

DON'T inhale dust – it may be injurious to health (see *Asbestos* heading).

DON'T allow any spilt oil or grease to remain on the floor – wipe it up straight away, before someone slips on it.

DON'T use ill-fitting spanners or other tools which may slip and cause injury.

DON'T attempt to lift a heavy component which may be beyond your capability – get assistance.

DON'T rush to finish a job, or take unverified short cuts.

DON'T allow children or animals in or around an unattended vehicle.

DON'T inflate a tyre to a pressure above the recommended maximum. Apart from overstressing the carcase and wheel rim, in extreme cases the tyre may blow off forcibly.

DO ensure that the machine is supported securely at all times. This is especially important when the machine is blocked up to aid wheel or fork removal.

DO take care when attempting to slacken a stubborn nut or bolt. It is generally better to pull on a spanner, rather than push, so that if slippage occurs you fall away from the machine rather than on to it.

DO wear eye protection when using power tools such as drill, sander, bench grinder etc.

DO use a barrier cream on your hands prior to undertaking dirty jobs – it will protect your skin from infection as well as making the dirt easier to remove afterwards; but make sure your hands aren't left slippery. Note that long-term contact with used engine oil can be a health hazard.

DO keep loose clothing (cuffs, tie etc) and long hair well out of the way of moving mechanical parts.

DO remove rings, wristwatch etc, before working on the vehicle – especially the electrical system.

DO keep your work area tidy – it is only too easy to fall over articles left lying around.

DO exercise caution when compressing springs for removal or installation. Ensure that the tension is applied and released in a controlled manner, using suitable tools which preclude the possibility of the spring escaping violently.

DO ensure that any lifting tackle used has a safe working load rating adequate for the job.

DO get someone to check periodically that all is well, when working alone on the vehicle.

DO carry out work in a logical sequence and check that everything is correctly assembled and tightened afterwards.

DO remember that your vehicle's safety affects that of yourself and others. If in doubt on any point, get specialist advice.

IF, in spite of following these precautions, you are unfortunate enough to injure yourself, seek medical attention as soon as possible.

Asbestos

Certain friction, insulating, sealing, and other products – such as brake linings, clutch linings, gaskets, etc – contain asbestos. *Extreme care must be taken to avoid inhalation of dust from such products since it is hazardous to health.* If in doubt, assume that they *do* contain asbestos.

Fire

Remember at all times that petrol (gasoline) is highly flammable. Never smoke, or have any kind of naked flame around, when working on the vehicle. But the risk does not end there – a spark caused by an electrical short-circuit, by two metal surfaces contacting each other, by careless use of tools, or even by static electricity built up in your body under certain conditions, can ignite petrol vapour, which in a confined space is highly explosive.

Always disconnect the battery earth (ground) terminal before working on any part of the fuel or electrical system, and never risk spilling fuel on to a hot engine or exhaust.

It is recommended that a fire extinguisher of a type suitable for fuel and electrical fires is kept handy in the garage or workplace at all times. Never try to extinguish a fuel or electrical fire with water.

Note: *Any reference to a 'torch' appearing in this manual should always be taken to mean a hand-held battery-operated electric lamp or flashlight. It does* **not** *mean a welding/gas torch or blowlamp.*

Fumes

Certain fumes are highly toxic and can quickly cause unconsciousness and even death if inhaled to any extent. Petrol (gasoline) vapour comes into this category, as do the vapours from certain solvents such as trichloroethylene. Any draining or pouring of such volatile fluids should be done in a well ventilated area.

When using cleaning fluids and solvents, read the instructions carefully. Never use materials from unmarked containers – they may give off poisonous vapours.

Never run the engine of a motor vehicle in an enclosed space such as a garage. Exhaust fumes contain carbon monoxide which is extremely poisonous; if you need to run the engine, always do so in the open air or at least have the rear of the vehicle outside the workplace.

The battery

Never cause a spark, or allow a naked light, near the vehicle's battery. It will normally be giving off a certain amount of hydrogen gas, which is highly explosive.

Always disconnect the battery earth (ground) terminal before working on the fuel or electrical systems.

If possible, loosen the filler plugs or cover when charging the battery from an external source. Do not charge at an excessive rate or the battery may burst.

Take care when topping up and when carrying the battery. The acid electrolyte, even when diluted, is very corrosive and should not be allowed to contact the eyes or skin.

If you ever need to prepare electrolyte yourself, always add the acid slowly to the water, and never the other way round. Protect against splashes by wearing rubber gloves and goggles.

Mains electricity and electrical equipment

When using an electric power tool, inspection light etc, always ensure that the appliance is correctly connected to its plug and that, where necessary, it is properly earthed (grounded). Do not use such appliances in damp conditions and, again, beware of creating a spark or applying excessive heat in the vicinity of fuel or fuel vapour. Also ensure that the appliances meet the relevant national safety standards.

Ignition HT voltage

A severe electric shock can result from touching certain parts of the ignition system, such as the HT leads, when the engine is running or being cranked, particularly if components are damp or the insulation is defective. Where an electronic ignition system is fitted, the HT voltage is much higher and could prove fatal.

Ordering spare parts

When ordering spare parts for any Suzuki, it is advisable to deal direct with an official Suzuki agent who should be able to supply most of the parts ex-stock. Parts cannot be obtained from Suzuki direct and all orders must be routed via an approved agent even if the parts required are not held in stock. Always quote the engine and frame numbers in full, especially if parts are required for earlier models.

The frame and engine numbers are stamped on a Manufacturer's Plate riveted to the steering head on the left-hand side. The frame number is also stamped on the frame itself on the right-hand side of the steering head. The engine number is stamped on the upper crankcase.

Use only genuine Suzuki spares. Some pattern parts are available that are made in Japan and may be packed in similar looking packages. They should only be used if genuine parts are hard to obtain or in an emergency, for they do not normally last as long as genuine parts, even though there may be a price advantage.

Some of the more expendable parts such as spark plugs, bulbs, tyres, oils and greases etc., can be obtained from accessory shops and motor factors, who have convenient opening hours and can often be found not far from home. It is also possible to obtain parts on a Mail Order basis from a number of specialists who advertise regularly in the motorcycle magazines.

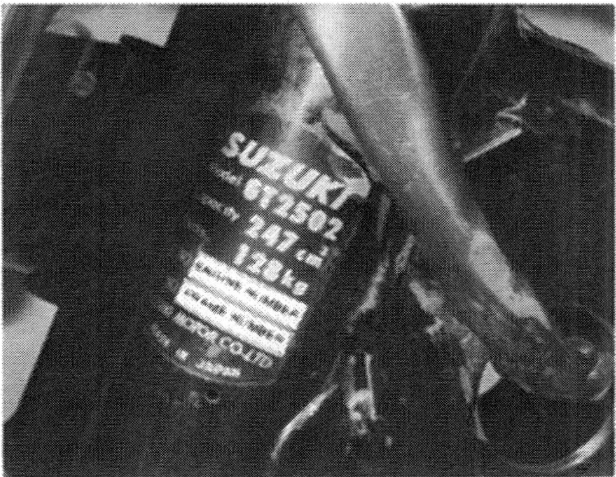

Machine identification plate location

Frame number location

Engine number location

Routine maintenance

Introduction

Periodic routine maintenance is a continuous process that commences immediately the machine is used. It must be carried out at specified mileage recordings, or on a calendar basis if the machine is not used frequently, whichever is the sooner. Maintenance should be regarded as an insurance policy, to help keep the machine in the peak of condition and to ensure long, trouble-free service. It has the additional benefit of giving early warning of any faults that may develop and will act as a regular safety check, to the obvious advantage of both rider and machine alike.

The various maintenance tasks are described under their respective mileage and calendar headings. Accompanying diagrams are provided, where necessary. It should be remembered that the interval between the various maintenance tasks serves only as a guide. As the machine gets older or is used under particularly adverse conditions, it would be advisable to reduce the period between each check.

For ease of reference each service operation is described in detail under the relevant heading. However, if further general information is required, it can be found within the manual under the pertinent section heading in the relevant Chapter.

In order that the routine maintenance tasks are carried out with as much ease as possible, it is essential that a good selection of general workshop tools are available.

Included in the kit must be a range of metric ring or combination spanners, a selection of crosshead screwdrivers and at least one pair of circlip pliers.

Additionally, owing to the extreme tightness of most casing screws on Japanese machines, an impact screwdriver, together with a choice of large or small crosshead screw bits, is absolutely indispensable. This is particularly so if the engine has not been dismantled since leaving the factory.

Weekly or every 200 miles (300 km)

1 *Topping up engine oil*

The oil tank level should be checked on a daily basis, prior to starting the engine. A small plastic window or sight glass gives an immediate visual warning of whether the oil level has dropped too low. Although it is quite safe to use the machine as long as oil is visible in the sight glass, it is recommended that the level is maintained to within about an inch of the tank filler neck, to allow a good reserve. It is advised that the tank is topped up at weekly intervals.

2 *Tyre pressures*

It is essential that the tyres are kept inflated to the correct pressure at all times. Under or over-inflated tyres can lead to accelerated rates of wear, and more importantly, can render the machine inherently unsafe. Whilst this may not be obvious during normal riding, it can become painfully and expensively so in an emergency situation, as the tyres' adhesion limits will be greatly reduced.

Check the tyre pressures with a pressure gauge that is known to be accurate. Always check the pressure when the tyres are cold. If the machine has travelled a number of miles, the tyres will have become hot and consequently the pressure will have increased. A false reading will therefore result.

It is recommended that a small pocket gauge is purchased and carried on the machine, as the readings on garage forecourt gauges can vary and may often be inaccurate.

The pressures given are those recommended for the tyres fitted as original equipment. If replacement tyres are purchased, the pressure settings may vary. Any reputable tyre distributor will be able to give this information.

Tyre pressure – X7 model

		Solo	with pillion
Front:			
	Normal	*21 psi (1.5 kg/cm²)*	*21 psi (1.5 kg/cm²)*
	High speed	*25 psi (1.75 kg/cm²)*	*25 psi (1.75 kg/cm²)*
Rear:			
	Normal	*25 psi (1.75 kg/cm²)*	*32 psi (2.25 kg/cm²)*
	High speed	*32 psi (2.25 kg/cm²)*	*35 psi (2.50 kg/cm²)*

Tyres pressures – X5 and SB200

	Solo	with pillion
Front	*25 psi (1.75 kg/cm²)*	*25 psi (1.75 kg/cm²)*
Rear	*28 psi (2.00 kg/cm²)*	*32 psi (2.25 kg/cm²)*

3 *Battery electrolyte level*

A Yuasa battery is fitted as standard. The battery is a lead-acid type and has a capacity of 5 or 8 amp hours.

The transparent plastic case of the battery permits the upper and lower levels of the electrolyte to be observed when the left-hand side panel has been removed. Maintenance is normally limited to keeping the electrolyte level between the prescribed upper and lower limits and by making sure that the vent pipe is not blocked. The lead plates and their separators can be seen through the transparent case, a further guide to the general condition of the battery.

Unless acid is spilt, as may occur if the machine falls over, the electrolyte should always be topped up with distilled water, to restore the correct level. If acid is spilt on any part of the machine, it should be neutralised with an alkali such as washing soda and washed away with plenty of water, otherwise serious corrosion will occur. Top up with sulphuric acid of the correct specific gravity (1.260 – 1.280) only when spillage has occurred. Check that the vent pipe is well clear of the frame tubes or any of the other cycle parts, for obvious reasons.

Sight glass provides means of checking oil tank level

4 *Control cable lubrication*

Apply a few drops of motor oil to the exposed inner portion of each control cable. This will prevent drying-up of the cables between the more thorough lubrication that should be carried out during the 2000 mile/3 monthly service.

5 *Rear chain lubrication and adjustment*

In order that the life of the rear chain can be extended as much as possible, regular lubrication and adjustment is essential.

Intermediate lubrication should take place at the weekly or 200 mile service interval with the chain in situ. Application of one of the proprietary chain greases contained in an aerosol can is ideal. Ordinary engine oil can be used, though owing to the speed with which it is flung off the rotating chain, its effectiveness is limited.

The chain lubricant may be applied via the inspection hole in the chain enclosure, where this is fitted.

Adjust the chain after lubrication, so that there is approximately 20 mm (¾ in) slack in the middle of the lower run. Always check with the chain at the tightest point as a chain rarely wears evenly during service.

Adjustment is accomplished after placing the machine on the centre stand and slackening the wheel nut, so that the wheel can be drawn backwards by means of the drawbolt adjusters in the fork ends.

Use aerosol chain lubricant between full lubrication intervals

Use pocket pressure gauge for regular tyre pressure checks

The torque arm nuts and the rear brake adjuster must also be slackened during this operation. Adjust the drawbolts an equal amount to preserve wheel alignment. The fork ends are clearly marked with a series of parallel lines above the adjusters, to provide a simple visual check.

6 *Safety check*

Give the machine a close visual inspection, checking for loose nuts and fittings, frayed control cables etc. Check the tyres for damage, especially splitting on the sidewalls. Remove any stones or other objects caught between the treads. This is particularly important on the front tyre, where rapid deflation due to penetration of the inner tube will almost certainly cause total loss of control.

7 *Legal check*

Ensure that the lights, horn and trafficators function correctly, also the speedometer.

Monthly or every 1000 miles (1500 km)

Complete all the checks listed in the weekly/200 mile service and then the following items:

1 *Oil pump adjustment*

The oil pump adjustment should be checked once each month to ensure that the correct amount of oil is being delivered to the engine. The check can be effected after removing the rear portion of the left-hand outer casing, and the oil pump cover. Check that the alignment marks on the pump body and operating lever coincide when the throttle is fully opened. See Chapter 2, Section 15 for further details.

2 *Checking the sparking plug*

Pull off the sparking plug cap and using the correct size plug spanner, remove the plug. Clean off any carbon or oil from the electrodes and using a feeler gauge, check the gap. Reset the gap, if necessary, after referring to Chapter 3. Refit the sparking plug into the cylinder head but do not overtighten it, as stripping of the threads could result. Refit the plug cap.

A new sparking plug should be fitted every 5000 miles (8000 km), or earlier if it is excessively worn, burnt or damaged.

The sparking plug gap should be set to 0.6–0.8 mm (0.024–0.031 in).

3 *Adjust slow running speed*

Adjust the slow running speed only if necessary. Refer to Chapter 2, Section 8.

4 *Clean the fuel tap filter*

Ensure the fuel tap filter is clean so that a smooth flow of fuel passes through the fuel tap. Turn the fuel tap to the OFF position. Using a 10 mm spanner, remove the cup and O-ring. Remove the filter gauze and wash it in clean petrol. Refit the filter O-ring and cup. Tighten the cup, using a 10 mm spanner.

5 *Exhaust system mountings*

Check that all exhaust mounting nuts and bolts are securely fastened. Any leaks will mean that the relevant gasket must be renewed.

6 *Clutch adjustment*

The clutch is correctly adjusted when the cable free play, measured between the butt end of the lever and its stop, is 4 mm (0.16 in) on X5 and SB200 models, and 2 – 3 mm (0.08 – 0.12 in) on X7 models. The clutch is adjusted as follows.

Slacken fully both cable adjuster locknuts and screw in the adjusters (one, lower, only on SB200 models) to gain the maximum cable free play. Remove the rubber bung (X7) or the inspection cover retained by three screws (X5 and SB200) from the left-hand engine cover to expose the clutch adjuster. Slacken the locknut and screw in the adjusting screw until resistance is felt. Unscrew the adjusting screw by ¼ – ½ turn and hold it in that position while tightening the locknut securely. Use the cable adjusters to set the correct cable free play, tighten all adjuster locknuts and refit the rubber bung or inspection cover.

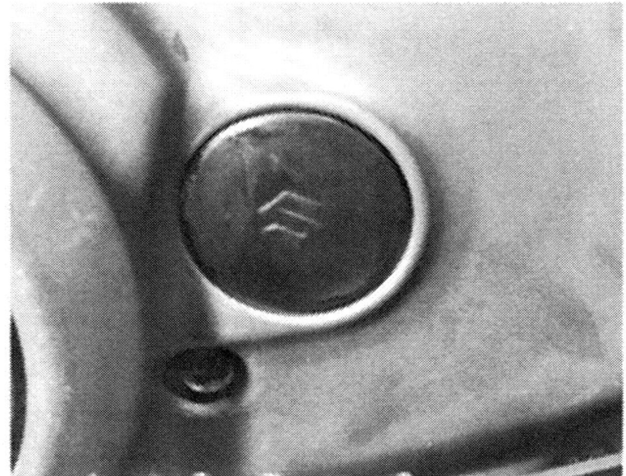
Clutch adjuster is covered by rubber bung on X7 model

7 *Checking the disc brake system*

Place the machine on its centre stand with the handlebars straight. Check the hydraulic fluid level in the master cylinder reservoir by viewing the fluid level through its translucent casing. If the fluid level is at or near the lower level limit, bring it up to the upper limit by adding a good quality hydraulic fluid meeting SAE J1703, DOT 3 or DOT 4 specification. Any marked drop in level should be viewed with suspicion as this is indicative of leakage somewhere in the system.

Check the hoses, pipes, connections, caliper and master cylinder for any sign of fluid leakage. If problems are apparent in this area refer to Chapter 5 for details of overhauling the system. Pad wear can be checked by viewing them from the front of the caliper. If worn to the red wear limit line, new pads should be fitted as described in Chapter 5, Section 4.

Fluid level must not fall below minimum line

8 *Adjusting drum brakes*

The front brake on SB 200 machines, and the rear brake on all models, is of the single leading shoe (sls) drum type. Both front and rear units should be adjusted so that brake operation commences just after the lever or pedal has commenced movement. As a rough guide, the front brake lever end should be 20–30 mm from the handlebar grip as the brake begins to take effect. If necessary, adjust the front brake cable at its lower end, by turning the adjusting nut.

The rear brake should start taking effect when the brake pedal has been depressed by about 20–30 mm. Adjustment can be carried out using the nut at the end of the brake operating rod.

Three monthly or every 2000 miles (3000 km)

Complete all the checks listed in the weekly/200 mile and monthly/1000 mile services, and then the following additional tasks.

1 *Cylinder head nuts – check and tighten*

Check and if necessary tighten the cylinder head nuts. The torque setting should be 14.5–18 lbf ft (2.0–2.5 kgf m).

Red line denotes maximum wear limit

2 *Clean the air filter*

If the air cleaner filter becomes blocked, intake resistance

Drum brake has wear range marked thus.

increases, resulting in loss of power and an increase in fuel consumption.

Release the left-hand cover from the air filter housing, and remove the air filter element. The element should be carefully washed out, dried and re-impregnated with engine oil. Full details of this procedure will be found in Chapter 2, Section 11.

3 *Carburettor and oil pump settings*

Referring to Chapter 2, Sections 8, 9 and 15, check the carburettor settings, and adjust them where necessary. Note that adjustment of the throttle cable free play will affect oil pump adjustment, so this should be checked. Do not make alterations to the setting unless this is strictly necessary – carburettors are generally best left alone if they are working efficiently.

4 *Gearbox oil change*

Warm the engine thoroughly to assist draining, then place a bowl or drain tray beneath the crankcase. Slacken and remove the hexagon headed drain plug on the underside of the crankcase, and allow the old oil to drain thoroughly. Clean and refit the drain plug, noting that the plug on the X7 doubles as a gearbox neutral detent plunger, and must always be fitted with the correct sealing washer.

Top up with good quality SAE 20W/40 engine oil, via the filter plug on the right-hand outer casing. Run the engine for a few minutes, then remove the hexagon-headed level screw from the right-hand outer casing (identified by sealing washer beneath its head) and top up, if necessary, to bring the oil level up to this hole.

Gearbox/primary drive oil capacity:

X7	*800 cc (1.69 US pint, 1.41 Imp pint)*
X5, SB 200	*700 cc (1.4 US pint, 1.2 Imp pint)*

5 *Checking the steering head bearings*

Place the machine on its centre stand so that the front wheel is raised clear of the ground. Turn the handlebars from lock to lock, noting any roughness which might indicate dry or damaged head races. If present, refer to Chapter 4, Section 3 for details on overhauling. Play in the steering head bearings will be evident if the bottom of the forks is grasped and pushed to and fro. Any detectable play will require adjustment.

Slacken the steering stem top bolt. The adjustment nut can be turned using a C-spanner until all free play has **just** been taken up. Recheck for free play, and then tighten the steering stem top bolt.

6 *Removing and lubricating the final drive chain*

In addition to intermediate chain lubrication as described in the weekly/200 mile routine maintenance section, the final drive chain should be attended to more thoroughly at greater intervals.

Place the machine on the centre stand and remove the rear portion of the left-hand casing to gain access to the sprocket. Separate the chain by prising off the spring link and sliding the chain ends apart. The chain can then be run off the sprockets.

Wash the chain carefully in paraffin (kerosene) using a stiff brush to remove all traces of road dirt. The chain should now be rinsed in petrol (gasoline) and hung up to dry off, or blown dry with compressed air.

The cleaned chain should be checked for wear by measuring the amount of stretch which has taken place. Lay the chain lengthways in a straight line and compress it at each end to take up all play. Anchor one end and pull on the other end to extend the chain and take up all play in the other direction. If the chain extends by more than the distance between two of the rollers it should be renewed and a close examination of both the engine and rear wheel sprockets must be carried out, to check for wear and sprocket tooth damage. See Chapter 5.16 for details.

The chain must be lubricated after cleaning,by immersing it in a molten chain lubricant, such as Linklyfe or Chainguard and then hanging it up to drain. This will ensure good penetration of the lubricant between the pins and rollers, making it less likely to be thrown off when the chain is in motion.

Position the two ends of the chain on the rear wheel sprocket, insert the link, fit the side plate and secure with the spring clip. Note that the closed end of the spring clip must be fitted pointing in the direction of motion.

Adjust the chain tension as previously stated in the weekly/200 mile service.

The chain fitted to GT250 X7 models is not fitted with a spring link, the chain being endless. Removal can be accomplished only by detaching the rear wheel and swinging arm unit. As a result chain cleaning and lubrication must be carried out with the chain in position, unless the owner is willing to go to the trouble of a large amount of dismantling.

7 *Control cable lubrication*

Lubricate the control cables thoroughly with motor oil or an all-purpose oil. A good method of lubricating the cables is shown in the accompanying illustration, using a plasticine funnel. This method has the disadvantage that the cables usually need removing from the machine. An hydraulic cable oiler which pressurises the lubricant overcomes this problem. Do not lubricate nylon lined cables (which may have been fitted as replacements), as the oil may cause the nylon to swell, thereby causing total cable seizure.

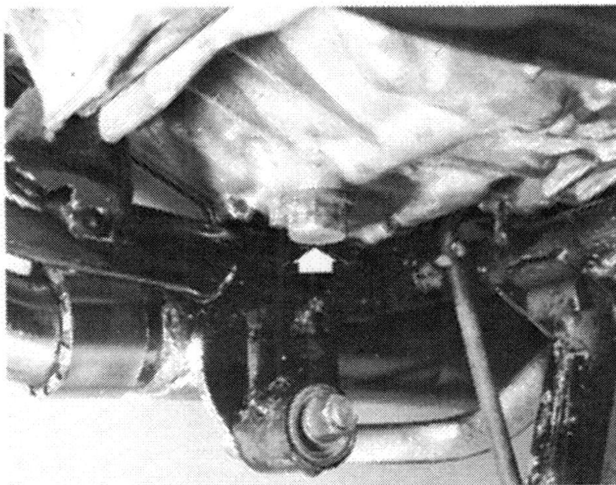
Gearbox drain plug (arrowed)

Refill gearbox as shown. Level plug (X7) arrowed

Slacken top bolt to permit steering head adjustment

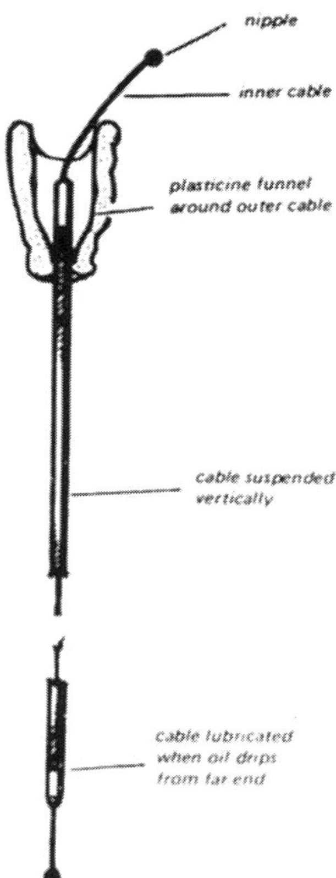

Control cable oiling

Six monthly or every 4000 miles (6000 km)

Complete all the checks listed in the weekly/200, monthly/1000 mile and three monthly/2000 mile services and then the following additional task:

1 Decarbonising the engine and exhaust system

To maintain peak performance, it is necessary to keep the cylinder head and the exhaust system free from accumulations of carbon. This is particularly important where the machine tends to be used for frequent short journeys, as this tends to accelerate the rate at which carbon builds up on these parts.

To effect this operation, it will be necessary to remove the cylinder head(s), releasing the eight nuts in a diagonal sequence. Carefully scrape off any carbon deposits from the cylinder head(s), and piston crowns, taking care not to score the soft alloy material of these parts. Carbon deposits should be removed from the exhaust ports in a similar manner after removal of the exhaust pipes to gain access. Attend to each cylinder separately, with the piston of that cylinder at BDC. Before refitting the head(s), remove any debris from the bores. Note that a new cylinder head gasket should be used. When tightening the cylinder head nuts, note the tightening sequence and torque settings given in Chapter 1.

The silencer baffles should be detached and cleaned by wire brushing. The baffles can be withdrawn after releasing their securing screws on the underside of the silencer. It may prove necessary to exert some force when removing them, as there is a tendency for them to become gummed into position.

Twelve monthly or every 8000 miles (12 000 km)

Complete all the checks listed under the weekly, monthly, three and six monthly headings, but only if they are not directly connected with the tasks listed below. More extensive dismantling is required when undertaking these latter tasks and reference to the relevant Chapters and Sections will be necessary in each case:

1 Dismantle, clean, examine and reassemble the carburettors.
2 Remove both wheels, grease the bearings and brake operating cams.
3 Check and grease the steering head bearings.

Quick glance maintenance adjustments and capacities

Engine (oil tank)	Fill to within 1 in of filler neck. Do not allow oil to fall to sight glass level or below	
Gearbox		
250 X7	800 cc (1.41 Imp pint 1.69 US pint)	
200 X5 and SB200	700 cc (1.2 Imp pint, 1.4 US pint)	
Sparking plug gap	0.6–0.7 mm (0.024–0.031 in)	
Tyre pressures		
	solo	with pillion
X7 – Front:		
Normal	21 psi (1.5 kg/cm^2)	21 psi (1.5 kg/cm^2)
High speed	25 psi (1.75 kg/cm^2)	25 psi (1.75 kg/cm^2)
X7 – Rear:		
Normal	25 psi (1.75 kg/cm^2)	32 psi (2.25 kg/cm^2)
High speed	32 psi (2.25 kg/cm^2)	35 psi (2.50 kg/cm^2)
X5 and SB200:		
Front	25 psi (1.75 kg/cm^2)	25 psi (1.75 kg/cm^2)
Rear	28 psi (2.00 kg/cm^2)	32 psi (2.25 kg/cm^2)

Recommended lubricants

Component	Type and specification
Engine	Good quality two-stroke oil, Suzuki CCI or similar
Gearbox	SAE 20W/40 engine oil
Front forks	SAE 10W/20 engine oil
Steering head bearings	High melting point grease
Wheel bearings	High melting point grease
Brake pivots	High melting point grease
Electrical contacts	WD40 or similar water-dispersant spray
Chain	Aerosol chain lubricant and hot-immersion lubricant; Linklyfe, Chainguard or similar

Working conditions and tools

When a major overhaul is contemplated, it is important that a clean, well-lit working space is available, equipped with a workbench and vice, and with space for laying out or storing the dismantled assemblies in an orderly manner where they are unlikely to be disturbed. The use of a good workshop will give the satisfaction of work done in comfort and without haste, where there is little chance of the machine being dismantled and reassembled in anything other than clean surroundings. Unfortunately, these ideal working conditions are not always practicable and under these latter circumstances when improvisation is called for, extra care and time will be needed.

The other essential requirement is a comprehensive set of good quality tools. Quality is of prime importance since cheap tools will prove expensive in the long run if they slip or break when in use, causing personal injury or expensive damage to the component being worked on. A good quality tool will last a long time, and more than justify the cost.

For practically all tools, a tool factor is the best source since he will have a very comprehensive range compared with the average garage or accessory shop. Having said that, accessory shops often offer excellent quality tools at discount prices, so it pays to shop around. There are plenty of tools around at reasonable prices, but always aim to purchase items which meet the relevant national safety standards. If in doubt, seek the advice of the shop proprietor or manager before making a purchase.

The basis of any tool kit is a set of open-ended spanners, which can be used on almost any part of the machine to which there is reasonable access. A set of ring spanners makes a useful addition, since they can be used on nuts that are very tight or where access is restricted. Where the cost has to be kept within reasonable bounds, a compromise can be effected with a set of combination spanners – open-ended at one end and having a ring of the same size on the other end. Socket spanners may also be considered a good investment, a basic 3/8 in or 1/2 in drive kit comprising a ratchet handle and a small number of socket heads, if money is limited. Additional sockets can be purchased, as and when they are required. Provided they are slim in profile, sockets will reach nuts or bolts that are deeply recessed. When purchasing spanners of any kind, make sure the correct size standard is purchased. Almost all machines manufactured outside the UK and the USA have metric nuts and bolts, whilst those produced in Britain have BSF or BSW sizes. The standard used in USA is AF, which is also found on some of the later British machines. Others tools that should be included in the kit are a range of crosshead screwdrivers, a pair of pliers and a hammer.

When considering the purchase of tools, it should be remembered that by carrying out the work oneself, a large proportion of the normal repair cost, made up by labour charges, will be saved. The economy made on even a minor overhaul will go a long way towards the improvement of a toolkit.

In addition to the basic tool kit, certain additional tools can prove invaluable when they are close to hand, to help speed up a multitude of repetitive jobs. For example, an impact screwdriver will ease the removal of screws that have been tightened by a similar tool, during assembly, without a risk of damaging the screw heads. And, of course, it can be used again to retighten the screws, to ensure an oil or airtight seal results. Circlip pliers have their uses too, since gear pinions, shafts and similar components are frequently retained by circlips that are not too easily displaced by a screwdriver. There are two types of circlip pliers, one for internal and one for external circlips. They may also have straight or right-angled jaws.

One of the most useful of all tools is the torque wrench, a form of spanner that can be adjusted to slip when a measured amount of force is applied to any bolt or nut. Torque wrench settings are given in almost every modern workshop or service manual, where the extent to which a complex component, such as a cylinder head, can be tightened without fear of distortion or leakage. The tightening of bearing caps is yet another example. Overtightening will stretch or even break bolts, necessitating extra work to extract the broken portions.

As may be expected, the more sophisticated the machine, the greater is the number of tools likely to be required if it is to be kept in first class condition by the home mechanic. Unfortunately there are certain jobs which cannot be accomplished successfully without the correct equipment and although there is invariably a specialist who will undertake the work for a fee, the home mechanic will have to dig more deeply in his pocket for the purchase of similar equipment if he does not wish to employ the services of others. Here a word of caution is necessary, since some of these jobs are best left to the expert. Although an electrical multimeter of the AVO type will prove helpful in tracing electrical faults, in inexperienced hands it may irrevocably damage some of the electrical components if a test current is passed through them in the wrong direction. This can apply to the synchronisation of twin or multiple carburettors too, where a certain amount of expertise is needed when setting them up with vacuum gauges. These are, however, exceptions. Some instruments, such as a strobe lamp, are virtually essential when checking the timing of a machine powered by CDI ignition system. In short, do not purchase any of these special items unless you have the experience to use them correctly.

Although this manual shows how components can be removed and replaced without the use of special service tools (unless absolutely essential), it is worthwhile giving consideration to the purchase of the more commonly used tools if the machine is regarded as a long term purchase Whilst the alternative methods suggested will remove and replace parts without risk of damage, the use of the special tools recommended and sold by the manufacturer will invariably save time.

Chapter 1 Engine, clutch and gearbox

Contents

Specifications

Engine

	GT250 X7	GT200 X5	SB200
Type	Twin cylinder air-cooled two-stroke		
Bore	54 mm (2.126 in)	50 mm (1.969 in)	50 mm (1.969 in)
Stroke	54 mm (2.126 in)	50 mm (1.969 in)	50 mm (1.969 in)
Capacity	247 cc (15.1 cu in)	196 cc (12.0 cu in)	196 cc (12.0 cu in)
Compression ratio	6.7:1	7.0:1	6.5:1
Lubrication	By engine driven pump, Suzuki CCI system		
Power output	–	–	18.3 hp @ 7500 rpm
Maximum torque	–	–	13.0 lbf ft @ 6000 rpm

Nominal dimensions and wear limits			
Cylinder bore diameter	54.000–54.015 mm (2.1260–2.1265 in)	50.000–50.015 mm (1.9685–1.9691 in)	50.000–50.015 mm (1.9685–1.9691 in)
Wear limit		50.090 mm (1.9720 in)	50.090 mm (1.9720 in)
Piston diameter	53.955–53.970 mm (2.1242–2.1248 in)	49.955–49.970 mm (1.9667–1.9673 in)	49.955–49.907 mm (1.9667–1.9673 in)
Wear limit	–	49.880 mm (1.9638 in)	49.880 mm (1.9638 in)
Piston-cylinder clearance	0.040–0.050mm (0.0016–0.0019 in)	0.040–0.050 mm (0.0016–0.0019 in)	0.040–0.050 mm (0.0016–0.0019 in)
Wear limit	0.120 mm (0.0047 in)	0.120 mm (0.0047 in)	0.120 mm (0.0047 in)
Piston ring gaps	0.15–0.35 mm (0.006–0.014 in)	0.15–0.35 mm (0.006–0.014 in)	0.15–0.35 mm (0.006–0.014 in)
Wear limit	0.80 mm (0.031 in)	0.80 mm (0.031 in)	0.80 mm (0.031 in)
Piston ring-groove clearance	–	0.02–0.06 mm (0.001–0.002 in)	0.02 – 0.06 mm (0.001–0.002 in)
Gudgeon pin-bore clearance	0.002–0.011 mm (0.0001–0.0004 in)	0.002–0.011 mm (0.0001–0.0004 in)	0.002–0.011 mm (0.0001 – 0.0004 in)
Wear limit	0.080 mm (0.0031 in)	0.080 mm (0.0031 in)	0.080 mm (0.0031 in)
Gudgeon pin diameter	13.995–14.000 mm (0.5510–0.5512 in)	13.995–14.000 mm (0.5510–0.5512 in)	13.995–14.000 mm (0.5510–0.5512 in)
Wear limit	13.965 mm (0.5498 in)	13.980 mm (0.5504 in)	13.980 mm (0.5504 in)
Gudgeon pin bore diameter	13.998–14.006 mm (0.5511–0.5514 in)	13.998–14.006 mm (0.5511–0.5514 in)	13.998–14.006 mm (0.5511–0.5514 in)
Wear limit	14.045 mm (0.5530 in)	14.045 mm (0.5530 in)	14.045 mm (0.5530 in)
Connecting rod small end bore diameter	18.003–18.011 mm (0.7088–0.7091 in)	18.003–18.011 mm (0.7088–0.7091 in)	18.003–18.011 mm (0.7088–0.7091 in)
Wear limit	18.051 mm (0.7107 in)	18.040 mm (0.7102 in)	18.040 mm (0.7102 in)
Crankshaft runout	0.05 mm (0.002 in) maximum	0.05 mm (0.002 in) maximum	0.05 mm (0.002 in) maximum
Clutch			
Type		Wet, multiplate	
No. of plain plates	5	5	5
No. of friction plates	5	6	6
No. of springs	6	7	7
Friction plate thickness	3.4–3.6 mm (0.134–0.141 in)	2.9–3.1 mm (0.11–0.12 in)	2.9–3.1 mm (0.11–0.12 in)
Wear limit	3.1 mm (0.122 in)	2.6 mm (0.10 in)	2.6 mm (0.10 in)
Plain plate maximum warpage	0.1 mm (0.004 in)	0.1 mm (0.004 in)	0.1 mm (0.004 in)
Clutch spring length	36.5 mm (1.437 in) minimum	33.6 mm (1.32 in) maximum	33.6 mm (1.32 in) maximum
Gearbox			
Type	6 speed, constant mesh	5 speed, constant mesh	4 speed, constant mesh
Ratios:			
1st	2.500:1	2.750:1	2.750:1
2nd	1.625:1	1.750:1	1.647:1
3rd	1.210:1	1.250:1	1142:1
4th	1.000:1	1.000:1	0.875:1
5th	0.863:1	0.875:1	–
6th	0.782:1	–	–
Selector fork to groove clearance	0.1–0.3 mm (0.004–0.012 in)	0.1–0.3 mm (0.004–0.012 in)	0.1–0.3 mm (0.004–0.012 in)
Wear limit	0.5 mm (0.019 in)	0.5 mm (0.019 in)	0.5 mm (0.019 in)
Mainshaft cluster length		92.0–92.1 mm (3.62–3.63 in)	92.0–92.1 mm (3.62–3.63 in)
Primary drive			
Type	Helical gear	Helical gear	Helical gear
Reduction ratio:			
Early models	3.100:1 (62/20T)	2.905:1 (61/21T)	2.905:1 (61/21T)
X models	3.100:1 (62/20T)	2.905:1 (61/21T)	3.368:1 (64/19T)
Final drive			
Type	Chain	Chain	Chain
Reduction ratio:			
Early models	2.733:1 (41/15T)	2.916:1 (35/12T)	3.083:1 (37/12T)
X models	2.866:1 (43/15T)	2.916:1 (35/12T)	2.666:1 (32/12T)

Torque settings

	GT250 X7	GT200 X5, SB200
Cylinder head nuts	14.5–18.0 lbf ft (2.0–2.5 kgf m)	14.5–18.0 lbf ft (2.0–2.5 kgf m)
Flywheel rotor nut	32.5–50.5 lbf ft (4.5–7.0 kgf m)	40.0–47.0 lbf ft (5.5–6.5 kgf m)
Gearbox sprocket fastener:		
2 x 6 mm bolts	4.5 – 7.5 lbf ft (0.6 – 1.0 kgf m)	4.5 – 7.5 lbf ft (0.6 – 1.0 kgf m)
1 x large nut	N/App	29 – 43 lbf ft (4.0 – 6.0 kgf m)
Clutch centre nut	29.0–43.5 lbf ft (4.0–6.0 kgf m)	21.5–36.0 lbf ft (3.0–5.0 kgf m)
Crankshaft pinion nut	29.0–43.5 lbf ft (4.0–6.0 kgf m)	65.0–72.5 lbf ft (9.00–10.0 kgf m)
Engine mounting bolts	18.0–29.0 lbf ft (2.5–4.0 kgf m)	8 mm: 13.0–16.5 lbf ft (1.8–2.3 kgf m) 10 mm: 18.0–29.0 lbf ft (2.5–4.0 kgf m)

1 General description

The engine gearbox units used in the three models covered in this manual are of similar design, there being detail differences in the state of tune, capacity and fittings employed. The GT250 X7 model shares the twin-cylinder air cooled two-stroke engine with the other models, but sports a number of unique features. Most significant of these is a sophisticated induction system which employs the traditional piston-porting arrangement at lower speeds but is supplemented by a twin reed valve arrangement to produce maximum power at high engine speeds.

The GT200 X5 is of similar construction to the X7, but does not employ the reed valve arrangement. The X5 is the only machine of the three to employ an electric starter. Both the X5 and the X7 feature engine casings finished in matt grey.

The SB200 is the commuter model of the range, and the engine unit is in effect a detuned version of the X5 type, but without the electric starter. Unlike the two sportier machines, the SB200 is fitted with polished engine casings.

On all three machines, conventionally finned cylinder heads and barrels are used, these being fitted with rubber damping blocks to reduce engine noise. On the X7 unit, separate cylinder barrels are employed, whilst on the SB200 and X5 units the cylinder barrels are cast together as a block. The 'Ram-Air' system of ducted cooling, used on the earlier GT250 models has been abandoned.

The crankcase is arranged to split horizontally as a unit, the engine and gearbox being built in unit. Primary drive is by gear to the constant-mesh gearbox. In this area, the three models differ considerably. The SB200 has a simple four-speed gearbox, whilst the X5 and X7 have five and six-speed gearboxes respectively, to make better use of their less flexible engines.

Engine lubrication is provided by a pump fed system known as Posi-Force. Unlike earlier versions of this arrangement, where oil was pumped directly to the stressed areas of the engine, this is a simplified type in which the oil is simply injected into the inlet tract and distributed by convection to the various moving parts. The gearbox components are housed in a separate chamber and are lubricated by an oil bath arrangement.

2 Operations with the engine/gearbox in the frame

It is not necessary to remove the engine/gear unit from the frame unless the crankshaft assembly and/or the gearbox components require attention. Most operations can be accomplished with the engine in place, such as:

1 Removal and replacement of the cylinder heads
2 Removal and replacement of the cylinder barrels and pistons
3 Removal and replacement of the alternator assembly
4 Removal and replacement of the clutch and primary drive
5 Removal and replacement of the starter motor and clutch (X5)
6 Renewal of the kickstart return spring (X5 and SB200)
7 Removal and replacement of the oil pump

When several operations need to be undertaken simultaneously, it would probably be an advantage to remove the complete unit from the frame, a comparatively simple operation that should take approximately twenty minutes. This will give the advantage of better access and more working space.

3 Operations with engine/gearbox removed

1 Removal and replacement of the crankshaft oil seals
2 Removal and replacement of the crankshaft assembly
3 Removal and replacement of the gear clusters, selectors and gearbox main bearings
4 Renewal of the kickstart return spring (X7)

4 Method of engine/gearbox removal

As described previously, the engine and gearbox are built in unit and it is necessary to remove the unit complete in order to gain access to either. Separation of the crankcases is accomplished after the engine unit has been removed and refitting cannot take place until the crankcases have been reassembled.

5 Removing the engine/gearbox unit from the frame

1 Place the machine securely on its centre stand, choosing a position which affords comfortable working access to both sides. It is helpful, but by no means essential, to raise the machine about 1 to 1½ feet from the ground by placing it on a stout table or ramp. This will ease working considerably by reducing the amount of stooping necessary. Slacken the large hexagon-headed drain plug on the underside of the gearbox, ensuring that neutral has been selected first. Place a suitable flat dish or drain tray beneath the unit and remove the plug completely to allow the gearbox oil to drain. This process will be aided if the engine is warm.

2 It will be noted that the drain plug on X7 models incorporates a spring-loaded plunger arrangement. This is the neutral detent plunger, and it is important that this is refitted unless the engine unit is to be dismantled completely. Note that the sealing washer beneath the head of the detent bolt must be refitted, as this also acts as a spacer, and its omission would cause the selector drum to become jammed.

3 Check that the petrol tap is in the 'Off' position, then prise

off the petrol feed pipe from the tap outlet stub. On X5 and SB200 models, release the dual seat by removing the securing bolts from the underside of the seat, at the rear. On X7 models, unlock the seat and lift it upwards. Remove the single bolt which secures the petrol tank at the rear, and pull the tank rearwards and upwards to clear the frame (all models). Place the tank in a safe place to avoid any subsequent damage.

4 Disconnect the battery leads, and preferably, remove the battery, the avoid any possible short circuits during dismantling.

5 On X7 models, remove the air filter housing end cover after releasing the two small bolts which secure it. Release the wing nut at the centre of the housing, and withdraw the filter element. Unscrew the carburettor tops and withdraw the throttle valves and needles. These components may be left attached to the throttle cables and lodged around the frame tubes, clear of the engine unit.

6 Slacken the four hose clips which retain the carburettor bodies to their respective hoses. The connecting hoses between the air filter and carburettors can be displaced into the filter housing to provide clearance for the carburettors to be removed. Alternatively, release the air filter housing's two retaining screws to allow the housing to be manoeuvred clear whilst the instruments are worked clear. This operation calls for considerable patience and dexterity. Note the sequence of removal carefully, as reassembly will demand an exact reversal and is even more awkward.

7 Slacken and remove the gearchange lever pinch bolt and draw the lever off its splined shaft. Resist any temptation to lever between the lever and the casing as the latter is easily damaged. Slacken the cross-head screws which retain the sprocket cover (X5 and SB200) or the left-hand engine casing (X7) and remove the cover to expose the gearbox sprocket and drive chain. The cover can be left connected to the clutch cable and lodged on the frame. The gearbox sprocket is secured by a retaining plate and two bolts, the latter being locked by a double tab washer, or by a large nut. Bend back the locking tabs and remove the two bolts, or the nut. It should now be possible to pull the sprocket off together with the drive chain. The sprocket can now be disengaged from the chain and the latter left to hang round the engine mounting spacer. Note that it may be necessary to slacken the rear wheel spindle and release the chain tension adjusters to provide sufficient free play. On X5 and SB200 models, where a chain joining link is fitted, the chain may be separated before the sprocket is slid off the shaft.

8 Release the pressed steel oil pump cover to gain access to the pump control cable and feed pipe. Disengage the oil pump cable and feed pipe. Disengage the oil pump cable nipple from the operating lever on the pump body. The cable adjuster should now be unscrewed, and the cable lodged clear of the engine. Squeeze the oil pump feed pipe from the oil tank to block the flow of oil, then prise the pipe off the stub at the pump. The oil in the tank can either be drained into a clean receptacle, or the pipe plugged with a suitable bolt or dowel to prevent leakage. Pull the pipe clear of the engine and lodge it against a frame tube.

9 Trace the output leads from the alternator up to the connector blocks, which should be separated. Note that the leads are colour coded and the connectors assymetrical, so there is no need to make specific notes on the connections involved. Pull the cables clear of the frame.

10 Release the exhaust pipe retaining plates at the exhaust ports, by removing the two bolts which secure each plate. Slacken the clamps where the exhaust pipes and silencers join, and pull the exhaust pipe clear. The silencers are best removed to permit easier access, after releasing the mounting nuts and bolts.

11 Moving to the right-hand side of the unit, disconnect the rear brake light switch operating spring. Slacken the gland nut which secures the tachometer drive cable to the right-hand engine casing, and pull the cable free. Pull off the sparking plug caps, noting which is left and right to avoid confusion during reassembly. It is helpful to mark the caps to this end.

12 Slacken and remove the engine mounting bolts to allow the engine unit to rest on the frame cradle. The engine front plates should be removed to give maximum clearance. Note that on X7 models a spacer is fitted between the frame and the rear left-hand side of the engine unit. This must be knocked clear before the engine unit can be removed. The engine unit is quite a tight fit in the frame, and consequently it is most useful to

5.3 A: Petrol tank mounting bolt B: Battery terminal

5.6 Carburettors should be removed as a pair

5.7a Gearchange pedal is secured by a pinch bolt

5.7b Release bolts and retainer plate to free sprocket

5.8 Disconnect oil pump cable from operating lever

5.10a Exhaust pipes are secured by two nuts and retainer

5.10b Silencer is retained to frame by a single bolt

5.11 Release gland nut to free tachometer drive cable

5.12 Engine/gearbox unit is now ready for removal

have an extra pair of hands to help manoeuvre it clear. The unit can be withdrawn from either side, but it will be easiest to remove it from the right-hand side if the kickstart lever is still attached.

6 Dismantling the engine and gearbox: general

1 Before commencing work on the engine unit, the external surfaces must be cleaned thoroughly. A motor cycle engine has very little protection from road grit and other foreign matter, which will sooner or later find its way into the dismantled engine if this simple precaution is not observed.
2 One of the proprietary engine cleaning compounds such as "Gunk" or "Jizer" can be used to good effect, especially if the compound is allowed to penetrate the film of oil and grease before it is washed away. When washing down, make sure that water cannot enter the inlet or exhaust ports or the electrical system, particularly if these parts are now more exposed.
3 Never use force to remove any stubborn part, unless mention is made of this requirement in the text. There is invariably good reason why a part is difficult to remove, often because the dismantling operation has been tackled in the wrong sequence.
4 Dismantling will be made easier if a simple engine stand is constructed that will correspond with the engine mounting points. This arrangement will permit the complete unit to be clamped rigidly to the work bench, leaving both hands free for the dismantling operation.

7 Dismantling the engine and gearbox: removing the cylinder head and barrels

1 If the cylinder head and barrels are to be removed with the engine unit in the frame, it should be noted that it will first be necessary to remove the carburettors and exhaust pires. See Section 5 paragraphs 6 and 10 for details. Note that for normal decarbonising work it is necessary to remove the cylinder head only, and the above steps may be omitted.
2 The cylinder head is retained by a total of eight studs and nuts, and these should be slackened progressively and in a diagonal sequence to avoid any risk of warpage. Lift the cylinder head clear, and place it to one side. The cylinder barrels, or cylinder block in the case of the X5 and SB200, can be slid upwards. Before removing the barrel(s) completely, stuff some clean rag in the crankcase mouths to prevent the ingress of debris or portions of broken piston ring. Note that on the X7 model, the barrels carry the reed valve, the stopper plates of which are easily damaged. For this reason, the barrels should be placed with the valve uppermost.

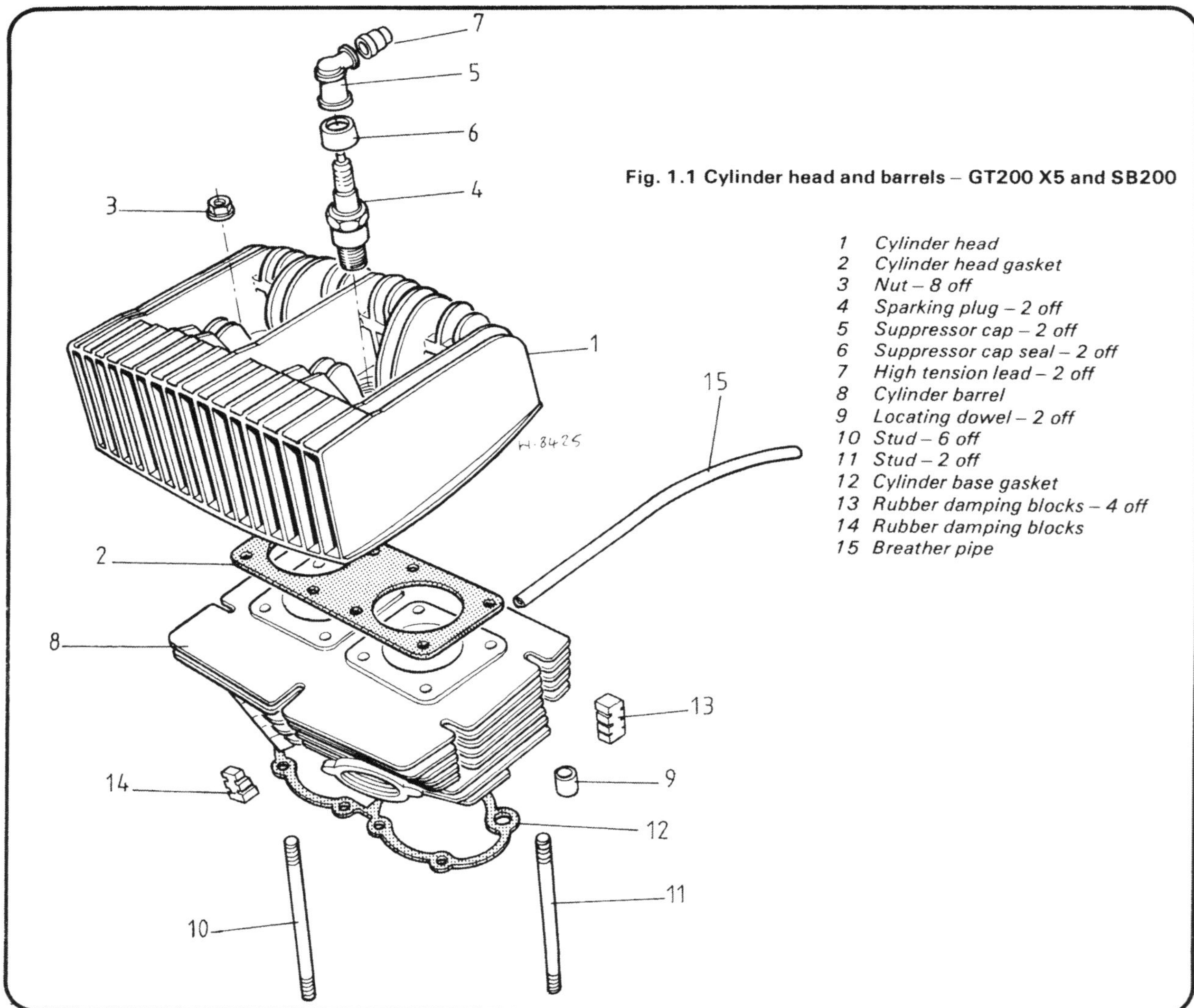

Fig. 1.1 Cylinder head and barrels – GT200 X5 and SB200

1 *Cylinder head*
2 *Cylinder head gasket*
3 *Nut – 8 off*
4 *Sparking plug – 2 off*
5 *Suppressor cap – 2 off*
6 *Suppressor cap seal – 2 off*
7 *High tension lead – 2 off*
8 *Cylinder barrel*
9 *Locating dowel – 2 off*
10 *Stud – 6 off*
11 *Stud – 2 off*
12 *Cylinder base gasket*
13 *Rubber damping blocks – 4 off*
14 *Rubber damping blocks*
15 *Breather pipe*

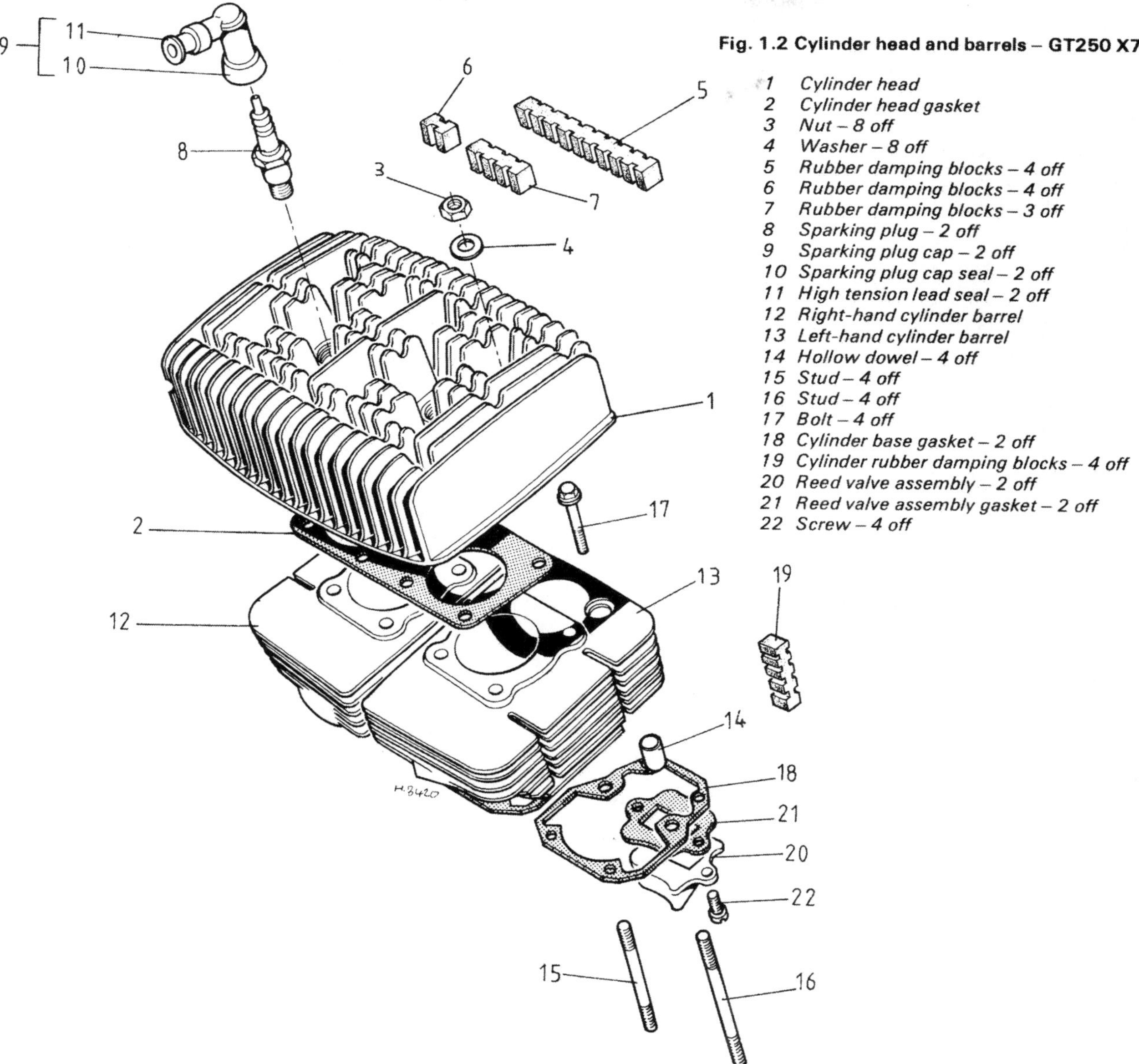

Fig. 1.2 Cylinder head and barrels – GT250 X7

1 Cylinder head
2 Cylinder head gasket
3 Nut – 8 off
4 Washer – 8 off
5 Rubber damping blocks – 4 off
6 Rubber damping blocks – 4 off
7 Rubber damping blocks – 3 off
8 Sparking plug – 2 off
9 Sparking plug cap – 2 off
10 Sparking plug cap seal – 2 off
11 High tension lead seal – 2 off
12 Right-hand cylinder barrel
13 Left-hand cylinder barrel
14 Hollow dowel – 4 off
15 Stud – 4 off
16 Stud – 4 off
17 Bolt – 4 off
18 Cylinder base gasket – 2 off
19 Cylinder rubber damping blocks – 4 off
20 Reed valve assembly – 2 off
21 Reed valve assembly gasket – 2 off
22 Screw – 4 off

8 Dismantling the engine and gearbox: removing the pistons and piston rings

1 Remove both circlips from each piston boss and discard them. Circlips should never be re-used if risk of displacement is to be obviated.

2 Using a drift of the correct diameter, tap each gudgeon pin out of the piston bosses until the piston complete with rings can be lifted off the connecting rod. Make sure the piston is properly supported during this operation, to prevent the connecting rod from bending.

3 If the gudgeon pin is a tight fit, the piston should first be warmed in order to expand the gudgeon pin bosses. A convenient way of warming the piston is to place a rag soaked in hot water on the crown.

4 When the pistons have been detached from the connecting rods, mark them on the inside of the skirt so that they will be replaced in identical positions. There is no need to mark the back and front because each piston has an arrow cast in the crown, which must always face the front of the machine.

5 The small end bearings take the form of caged needle rollers. Each roller assembly will lift out of the connecting rod eye.

6 Note that the piston rings are pegged so that they will remain in a set location. This is important, otherwise the rings will rotate whilst the engine is running, permitting the ends to become trapped in the ports and broken.

7 To remove the rings spread the ends sufficiently with the thumbs to allow each ring to be lifted clear of the piston. This is a very delicate operation which must be handled with great care. Piston rings are brittle and they break very easily.

8 If these rings are stuck in their grooves or have become gummed by oily deposits, it is sometimes possible to free them by working small strips of tin along the back, to give a 'peeling' action.

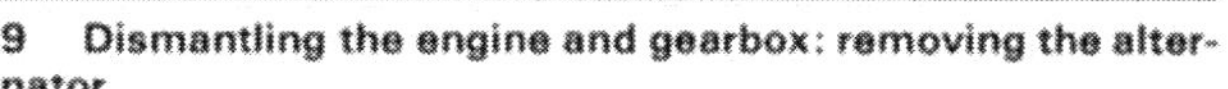
8.4 Pistons should be marked left and right during removal

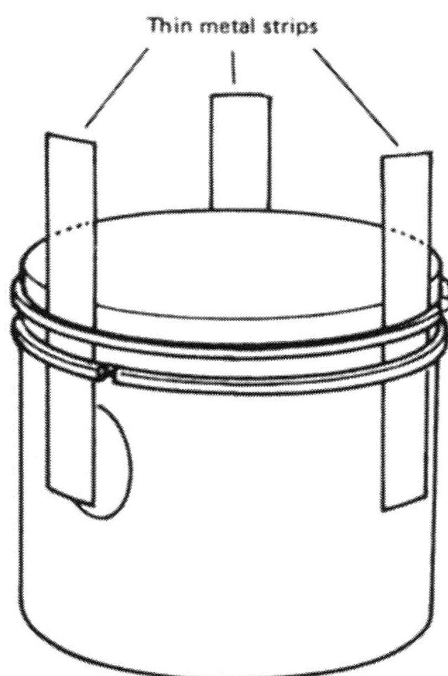

Fig. 1.3 Freeing gummed rings

9 Dismantling the engine and gearbox: removing the alternator

1 If the alternator is to be removed with the engine unit installed in the frame, it will be necessary to remove the front section of the left-hand engine casing (the complete left-hand engine casing in the case of X7) to gain access to the assembly. On all three models, electronic ignition is fitted, and the ignition pickup is an external appendage of the alternator stator in each case. The alternator of the X7 is conventionally arranged, having the stator screwed to the crankcase and the rotor keyed to the crankshaft, outboard of the stator. The X5 rotor incorporates the starter clutch and driven pinion, and consequently, its rotor faces outwards with the stator mounted inside the outer cover. The arrangements used on the SB200 is similar to that employed on the X5.

2 In each case, it will be necessary to prevent the crankshaft from rotating whilst the rotor securing nut is released. If this is to be done with the engine in situ, top gear should be selected and the rear brake applied, thus locking the engine and transmission. With the engine on the bench, a strap or chain wrench can be used around the rotor, as an alternative to the manufacturer's rotor holding tool, provided that care is taken not to damage the electronic ignition pickup, where this is still present. If the cylinder head and barrel have been removed, a bar can be passed through one of the connecting rod eyes, and arranged to rest on wooden blocks placed at the crankcase mouth. This will allow the nut to be slackened without fear of marring the gasket face.

3 Slacken and remove the rotor securing nut. In the absence of the manufacturer's slide hammer arrangement, some alternative method of drawing the rotor off its taper must be devised. A conventional three-legged puller can be used to good effect, noting that on X7 models, there is little room between the rotor and stator in which the puller legs may be inserted, necessitating particularly thin jaws. Alternatively, a metal plate can be drilled to correspond with the three extraction holes on the rotor, and a central nut and bolt used to draw the rotor off. Whichever method is chosen, take care not to exert excessive force, as damage may easily be caused.

4 On X5 models, the starter clutch is attached to the rotor, and will draw off with it. On X7 models, remove the stator assembly by releasing the three securing screws, and place it to one side.

9.3 Some form of puller is necessary for rotor removal

10 Dismantling the engine and gearbox: removing the starter motor and idler pinion – X5 only

1 The GT200 X5 is the only model of the three to be fitted with an electric starter. The motor is housed in a compartment in the underside of the crankcase, and drives through an idler pinion to a gear and roller clutch arrangement mounted inboard of the alternator rotor. Before the starter motor and idler pinion can be removed, it is necessary to remove the alternator rotor and starter clutch unit as described in Section 9 of this Chapter.

2 With the rotor and starter clutch removed, the starter motor idler gear and shaft can be withdrawn from the casing, taking care not to lose the two thrust washers. Moving to the underside of the unit, release the four bolts which secure the starter motor cover, and lift the cover away. It will be noted that the motor is secured by one additional bolt beneath the cover, and when this has been released the motor can be withdrawn from its housing.

11 Dismantling the engine and gearbox: removing the clutch

1 Access to the clutch and primary drive components is gained after removing the kickstart lever and right-hand outer cover, the latter being secured by eleven cross-head screws. As the cover is drawn off, a small amount of residual transmission oil may be spilt, and some provision must be made to catch this. If the engine unit is still in place in the frame, and it is wished to remove the primary drive assembly complete, the crankshaft pinion nut should be slackened at this stage. Bend back the tab washer, select top gear and apply the rear brake to prevent the crankshaft from turning, then slacken the nut.
2 The clutch used on the X5 and SB200 differs from that fitted to the X7 in some respects. On the X5 and SB200 clutch, the pressure plate is retained by tension springs. These are anchored by small pins at the pressure plate. They can be removed by inserting a screwdriver in the protruding loop of each spring and levering the spring outwards to release pressure on the pin. The pin can then be withdrawn, using a pair of pointed-nose pliers, and the spring released. This operation should be repeated on the remaining five springs, and the pressure plate lifted away.
3 In the case of the X7, conventional compression springs are used, and these and the pressure plate can be released by unscrewing the six retaining screws. The screws should be slackened progressively, and in a diagonal sequence until pressure on each screw is relieved, the screws, washers and springs can then be removed, and the pressure plate lifted away. The clutch friction and plain plates can now be removed as a group and placed to one side. Remove the clutch pushrod and the caged roller thrust race from the clutch centre.
4 The clutch centre on all three models is secured by a large nut, and it will be necessary to contrive some method of holding the clutch centre whilst this is slackened. Suzuki provide various service tools for holding the clutch centre on the different models, and these may be ordered from Suzuki dealers. In view of the cost of these rather specialised items, they may not be considered worthwhile by the average owner, and some improvised method will be required as an alternative.
5 If the engine unit is in position in the frame, the gearbox can be immobilised by selecting top gear and applying the rear brake whilst the nut is removed. With the unit on the workbench, an inexpensive holding tool can be made up as described below.
6 Obtain a strip of $\frac{1}{8}$ in mild steel, or similar, about one foot long by one inch wide. Using a blowlamp, heat the last inch or two of the strip until it is cherry red, and then bend it at 90° to the main length. When the strip has cooled off, it can be secured to the side of the clutch centre using a large worm-drive hose clip. The arrangement is shown in use in Fig 1.6
7 With the clutch centre held as described above, knock back the tab washer, and slacken the securing nut. The clutch centre may now be removed, noting the thrust washer(s) fitted behind it, followed by the clutch outer drum, and its thrust washer, where fitted. Lay these parts out in order, to ensure accurate reassembly.

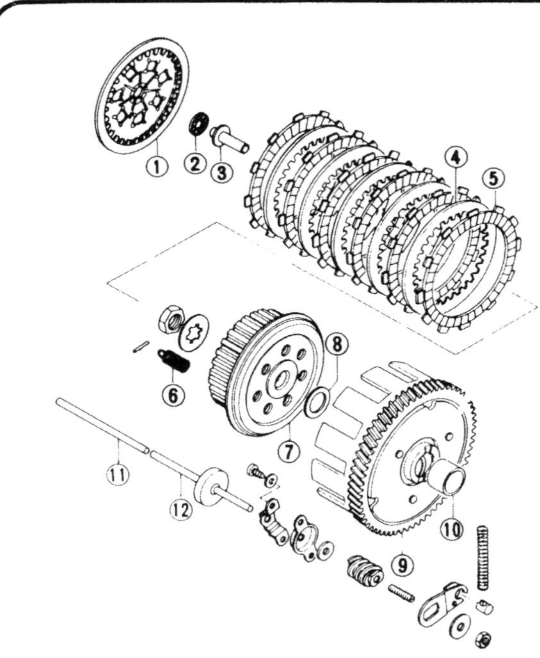

Fig. 1.4 Clutch assembly – GT 200 X5 and SB 200

1 *Clutch pressure plate*
2 *Thrust bearing*
3 *Clutch release piece*
4 *Clutch plate – plain – 5 off*
5 *Clutch plate – friction – 6 off*
6 *Clutch spring – 7 off*
7 *Clutch centre*
8 *Thrust washer*
9 *Primary driven gear*
10 *Spacer*
11 *Push rod*
12 *Push rod*

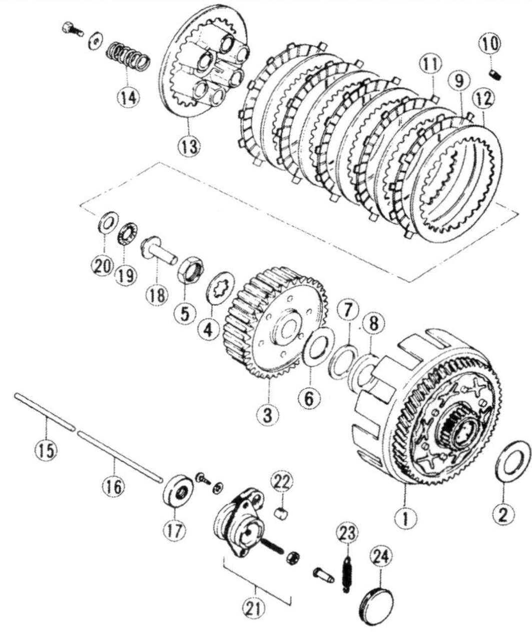

Fig. 1.5 Clutch assembly – GT 250 X7

1 *Primary driven gear assembly*
2 *Thrust washer*
3 *Clutch centre*
4 *Splined washer*
5 *Nut*
6 *Thrust washer*
7 *Rubber washer*
8 *Thrust washer*
9 *Inner friction plate*
10 *Damping blocks – 3 off*
11 *Friction plate – 4 off*
12 *Plain plate – 5 off*
13 *Clutch pressure plate*
14 *Clutch spring – 6 off*
15 *Push rod – short*
16 *Push rod – long*
17 *Push rod oil seal*
18 *Clutch release piece*
19 *Thrust bearing*
20 *Thrust washer*
21 *Release mechanism*
22 *Trunnion*
23 *Return spring*
24 *End cap*

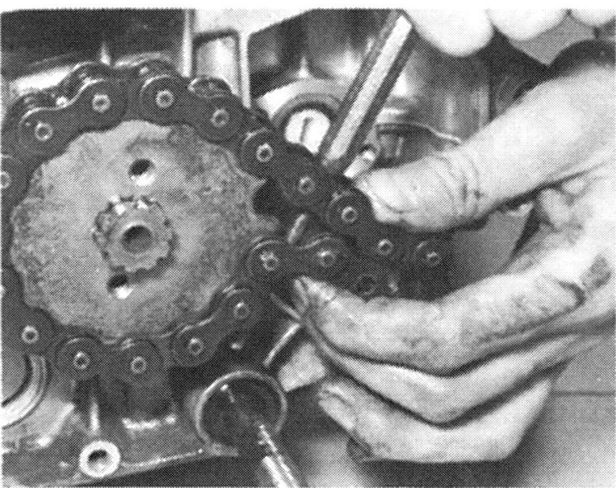

11.5 Gearbox can be immobilised as shown above

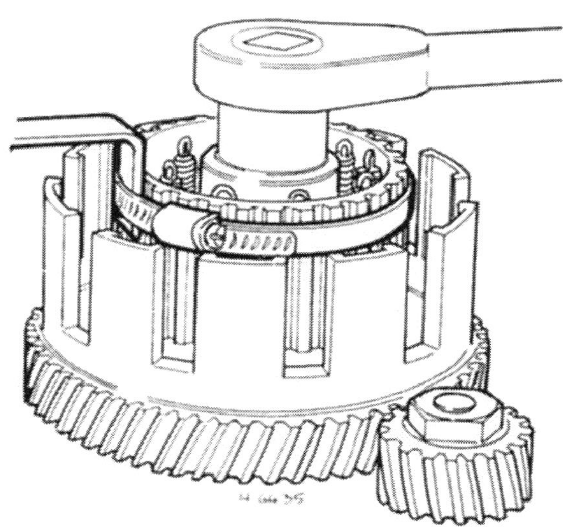

Fig. 1.6 Method of holding clutch centre whilst retaining nut is slackened

12 Dismantling the engine and gearbox: removing the crankshaft pinion(s)

1 If the clutch and primary drive only are to be removed, the crankshaft pinion nut should have been slackened prior to the removal of the clutch assembly, as described in Section 11, paragraph 1. If the pinion(s) are to be removed in the course of a complete engine stripdown, the crankshaft can be immobilised by passing a bar through one of the connecting rod eyes and supporting its ends on wooden blocks placed on the crankcase.

2 With the crankshaft held as described above, knock back the locking tab and slacken the retaining nut. The crankshaft pinion, and in the case of the X5, the tachometer drive pinion, can now be pulled off the crankshaft end. Do not omit to remove the Woodruff key and place it with the pinion(s) for safe keeping.

13 Dismantling the engine and gearbox: removing the kick-start mechanism

1 The kickstart mechanism can be reached after removing the right-hand outer cover and the clutch assembly, as described in the preceding Sections. The types of mechanism differ between models, and are therefore dealt with separately.

GT250 X7

2 Using circlip pliers, remove the circlip from the end of the kickstart idle pinion shaft, and remove the pinion. Slide off the white plastic tachometer drive pinion, followed by the kickstart pinion itself. Temporarily refit the kickstart lever, and turn it through about $\frac{1}{4}$ turn to disengage the ratchet segment from its stop. Slide the ratchet off its splines, then allow the shaft to unwind, thus relieving the return spring tension. No further dismantling can be undertaken at this stage, and it will be necessary to separate the crankcase halves to remove the kick-start shaft and return spring.

GT200 X5 and SB200 models

3 Displace the white plastic return spring guide using a screwdriver or similar to dislodge it from inside the spring coils. Grasp the spring end with a pair of pliers at the point where it enters the locating hole in the kickstart shaft. Disengage the spring end, allowing it to unwind in a controlled fashion. The spring can then be removed entirely. No further dismantling is possible at this stage, and it will be necessary to separate the crankcase halves to allow the rest of the mechanism to be removed.

14 Dismantling the engine and gearbox: removing the gear selector mechanism – X5 and SB200

1 Slacken the central screw on the end of the selector drum, and lift the end plate away. Remove the selector drum pins, noting that the neutral pin is shaped differently from the remainder. Remove the shouldered bolt on which the L-shaped stopper arm pivots. The stopper arm can be lifted away after disengaging its return spring. The gear selector mechanism and its operating shaft can now be withdrawn from the casing.

2 It will be noted that the large bearing retainer plate, to which the stopper arm spring was anchored, bridges the two halves of the crankcase. If the engine unit is to be dismantled further, the plate must be removed, using an impact driver to release the screws which retain it.

15 Dismantling the engine and gearbox: removing the gear selector mechanism – X7

1 The GT250 X7 uses an entirely different gear selector arrangement to that employed on the X5 and SB200 models. The gearchange shaft terminates in a toothed segment which engages corresponding teeth on the ratchet mechanism at the end of the selector drum. The gearchange shaft and toothed segment can be withdrawn from the casing and placed to one side.

2 The ratchet mechanism is located by a guide plate, which can be removed after its two retaining screws have been released. Opposite the guide plate is a second plate which is used to deflect the ratchet pawls when the assembly is turned. This too should be removed after releasing its retaining screws. The ratchet assembly can be withdrawn, taking care to hold the pawls together as it is pulled clear. Take great care not to lose the pawls, pins or springs. If the crankcase halves are to be separated, it will be necessary to remove the gearbox mainshaft bearing retainer, which bridges the crankcase halves. The two screws which retain it will probably demand the use of an impact driver to dislodge them.

16 Separating the crankcase halves: GT250 X7

1 With the engine unit stripped of all ancillary components, as described in the preceding Sections, arrange it on the workbench with the underside facing upwards. There are a total of thirteen bolts fitted from the underside, and these must be slackened in the reverse of the tightening sequence shown. To this end, refer to photograph 33.2b and slacken the bolts, commencing with 13 and working back to 1.
2 Turn the unit over again, and slacken the six remaining bolts which are fitted from the upper surface (see photograph 33.2c). The upper crankcase half can now be lifted away, leaving the crankshaft and gear clusters in the lower casing. It may well be discovered that the crankcase halves are reluctant to separate, and may require a certain amount of persuasion. It is permissible to tap around the joint using a soft-faced mallet, or a hammer and a hardwood block interposed between it and the casing. On no account should a metal hammer be used directly on the casing, as the soft aluminium is easily damaged by this method. Avoid any temptation to lever the joint apart with screwdrivers or other implements, as this will invariably damage the sealing faces.
3 With the upper crankcase half removed, the kickstart shaft and mechanism can be lifted away. Remove the crankshaft assembly, noting the half-ring retainer which locates the right-hand main bearing. The gearbox shafts are similarly located, and can now be removed and placed to one side.
4 It is now necessary to remove the selector drum and forks. The latter run on support rods which are fitted from the right-hand side of the casing. One of the rods is retained by the selector drum pawl guide plate, which has already been removed, whereas the rearmost rod is retained by the selector shaft centring spring locating pin. This pin has a hexagon machined on it, and should be unscrewed. The rear pin can now be withdrawn from the casing, using a pair of pointed-nose pliers to effect its removal. Initial resistance may be high, due to the oil film around the pin and the blind inner recess. Once any hydraulic lock of this nature is overcome, the rod will slide out easily.
5 As the rod is withdrawn, its two selector forks will be displaced, and these should be retrieved and refitted on the rod in their correct positions as an aid to reassembly. It will now be possible to reach the end of the selector drum detent spring, which is anchored on a small pin at the bottom of the casing. Using pointed-nose pliers, disengage the spring. The front selector fork rod can now be withdrawn, and the single selector fork and the selector drum detent arm removed.
6 Before the selector drum is removed, the neutral indicator switch should be released. Unscrew the two screws which secure the switch cover to the side of the crankcase, and lift it away. The small pin and spring in the end of the selector drum should be removed to preclude loss. The selector drum can now be displaced and withdrawn from the casing.

17 Separating the crankcase halves: GT200X5 and SB200

1 Arrange the partly-stripped unit on the workbench, with the crankcase mouths uppermost. Slacken and remove the three upper crankcase bolts, then invert the unit to gain access to the bolts on the underside.
2 The lower bolts, a total of thirteen, are numbered with the numerals cast into the casing, adjacent to the bolt holes. The numbering denotes the tightening sequence, and this should be reversed when removing the bolts, to avoid any risk of warping the casing halves.
3 Unlike the X7, the X5 and the SB200 units are dismantled by removing the lower casing half, leaving the internal components arranged in the inverted upper casing. Tap around the joint with a soft-faced mallet to help break the seal, then lift the lower casing half away. Do not attempt to force separation by the use of levers or metal hammers, as this will almost certainly cause damage to the casings.
4 Displace the kickstart shaft and mechanism and place it to one side. The two gear clusters can be removed as an assembly, noting the half-rings which locate the gearbox bearings in the casing. Lift the crankshaft assembly clear, again noting the half-rings which may remain with the bearings or may lodge in the casing grooves.
5 Working from the right-hand side of the upper casing, release the two bolts which secure the selector drum retainer plate and lift this away. The single cross-head screw which retains the front selector fork rod should also be removed. On the opposite end of the selector drum, dismantle the neutral indicator switch by removing the white plastic cover, followed by the small pin and spring from their recess in the selector drum end.
6 Grasp each of the selector fork rods in turn and slide them out of the casing towards the right-hand side. A pair of pointed-nose pliers will prove ideal for this operation. As the rods are removed, the selector forks will be displaced, and these should be retrieved from the casing and refitted on the rods in their respective positions. It will be noted that the X5 has three selector forks whilst the SB200 has two.
7 The selector drum can now be displaced and withdrawn from the casing after removing the detent assembly by unscrewing the hexagon-headed bolt on the outside of the casing, immediately adjacent to the drum. The various gearbox components should be laid out on a clean surface in their correct relative positions as an aid to reassembly.

18 Removing the crankshaft and gearbox main bearings

1 Before the crankshaft outer main bearings can be removed, it is first necessary to remove the outer oil seals. These are a push fit on the crankshaft and are quite easily withdrawn. A special puller is needed to remove the outer main bearings without risk of damage to the crankshaft assembly, although it is possible to use a sprocket puller with thin jaws since there is a thrust washer between each bearing and the nearest flywheel, which provides the necessary clearance.
2 Generally speaking, it is preferable to service exchange the entire crankshaft assembly if any of the bearings are suspect. If the two outer bearings need replacing, it is highly probable that the centre bearing will be in the same condition. It is quite beyond the means of the home mechanic, or for that matter, the majority of motor cycle repair specialists, to separate and realign the crankshaft assembly without the appopriate equipment. In consequence, the Suzuki Service Exchange Scheme will provide a completely reconditioned crankshaft assembly together with new main bearings, small end bearings and oil seals for a fixed sum, in exchange for the crankshaft that requires attention. Any accredited Suzuki dealer should be able to provide this facility, often ex-stock.
3 By way of explanation, the crankshaft assembly can be regarded as two quite independent flywheel assemblies, coupled together by a 'middle' crankshaft that presses into the inner flywheels. The middle crankshaft carries the centre main bearing and also the two innermost oil seals. The two separate flywheel assemblies are built up first, aligned and checked for runout. They are then coupled together by the middle crankshaft and the entire assembly is checked for runout at both ends. It follows that a very high standard of accuracy has to be observed, in order to preserve the smooth running of the engine.
4 The gearbox oil seals are located between the two casing halves, and are easily removed for examination. The mainshaft has a large diameter journal ball bearing at the left-hand or drive-side end and a caged needle roller bearing on the right-hand end. The reverse applies in the case of the layshaft; the journal ball bearing is on the right and the caged needle roller bearing on the left. There should be no difficulty in pulling these bearings off the shafts.

19 Examination and renovation: general

1 Before examining the parts of the dismantled engine unit for wear, it is essential that they should be cleaned thoroughly. Use a paraffin/petrol mix to remove all traces of old oil and sludge that may have accumulated within the engine.
2 Examine the crankcase castings for cracks or other signs of damage. If a crack is discovered, it will require professional repair.
3 Examine carefully each part to determine the extent of wear, checking with the tolerance figures listed in the Specifications section of this Chapter. If there is any question of doubt, play safe and renew.
4 Use a clean, lint-free rag for cleaning and drying the various components. This will obviate the risk of small particles obstructing the internal oilways, causing the lubrication system to fail.

20 Gearbox mainshaft and layshaft: dismantling and reassembly

1 The gearbox clusters should not be disturbed needlessly, and need only be stripped where careful examination of the whole assembly fails to resolve the source of a problem, or where obvious damage, such as stripped or chipped teeth is discovered.
2 The most significant problem facing owners of X5 and SB200 machines, is that the mainshaft 2nd gear pinion is a press-fit on the shaft, and its removal may cause a number of problems. The prescribed method of removal is to use an hydraulic press, which is almost certainly a facility unavailable to most owners. An alternative is to use a suitably thin-jawed legged puller or bearing extractor to draw the pinion off its splines. The splines should then be cleaned to remove any residual locking compound, and the remaining pinions removed in the normal way.
3 When reassembling the mainshaft, ensure that the end of the shaft and internal bore of the 2nd gear pinion are clean and dry. Coat the pinion bore with Suzuki Lock Super 1303B or an equivalent bearing and gear locking compound. When the pinion is fitted, it is essential to check that no excess locking fluid finds its way onto neighbouring components. The pinion can be refitted using a press or a tubular drift. It is important to ensure that the overall length of the assembled shaft, measuring from the outer faces of the 2nd and 1st gear pinions, is 92.0 - 92.1 mm (3.62 - 3.63 in), and this should be checked using a vernier caliper as the shaft is assembled. On SB200 models the measurement should be made from the outer face of the 2nd gear pinion and the outer face of the spacer on the opposite end of the shaft.
4 The procedure for removing the 2nd gear mainshaft pinion on the X7 model is somewhat less involved, but still requires a slightly unorthodox approach. The pinion is secured by a round section wire circlip which locates in a recess in the pinion boss. It will be seen that in its normal position, it is impossible to remove the pinion, as the circlip is masked by the pinion boss. To gain access to the circlip, it will first be necessary to release the circlip from the groove behind the 6th gear pinion, sliding it towards the centre of the shaft. The 6th and 2nd gear pinions can now be moved inwards, allowing the hidden clip to be displaced.
5 The remaining mainshaft components can be removed in a straightforward manner. As each item is slid off the shaft, it should be laid out on a clean surface in the exact order of removal. This will make reassembly much simpler. The layshaft on all models is dealt with in a similar fashion. When rebuilding the shafts, reference should be made to the accompanying line drawing(s) for details of the location and positioning of the gear pinions, washers and circlips. The manufacturer recommends that new circlips should always be used to preclude any possibility of a strained circlip failing in service. Washers should be fitted with their chamfered face towards the adjoining pinion. The six-speed gearbox of the X7 model is slightly less straightforward than that of the X5 and SB200, and to this end, reference should be made to the photographic sequence which accompanies this Section.
6 To facilitate reassembly, a brief summary of the rebuilding sequence for each model is given below:

SB 200

7 The mainshaft can be identified by its integral 1st gear pinion. Fit the 3rd gear pinion and retain it with a new circlip. Slide on the 4th gear pinion, with selector groove nearest the 3rd gear pinion. Clean the splines, and apply a high-shear locking fluid, before pressing the 2nd gear drive pinion into position. Check gear cluster length. Finally, fit the spacer to the end of the shaft. The layshaft is assembled as follows; fit the spacer, followed by the 2nd gear pinion, a second spacer (identical to first) and the 4th gear pinion, dogs outwards. This assembly should then be secured with a new circlip. Fit the 3rd gear pinion, dogs and selector groove inward, 1st gear pinion and, finally, the plain thrust washer.

GT200 X5

8 Fit the 4th gear pinion to the mainshaft, and secure it with a new circlip. Fit the 3rd gear pinion, selector groove inwards, and the 5th gear pinion. Clean the splines and apply a high-shear locking fluid, then press on the 2nd gear pinion, checking the overall length of the cluster. The layshaft is assembled in the following order. Fit the 2nd gear pinion, circlip, 5th gear pinion (selector groove outwards), circlip, washer, 3rd gear pinion, circlip, 4th gear pinion (selector groove inwards), 1st gear pinion and plain thrust washer.

GT250 X7

9 Fit the mainshaft 5th gear pinion, thrust washer, circlip and 3rd/4th gear combined pinion. Fit the circlip, but do not locate it in its groove. Position it close to the 3rd/4th gear pinion. Fit a thrust washer, 6th gear pinion, 2nd gear pinion and secure with round wire circlip. Slide 6th and 2nd gear pinions down to allow 6th gear pinion circlip to be fitted in its groove.
10 Fit the 2nd gear layshaft pinion, thrust washer and circlip to the gearbox sprocket end of the layshaft. Fit the 6th gear pinion, circlip, thrust washer, 3rd gear pinion, 4th gear pinion, thrust washer, circlip, 5th gear pinion, 1st gear pinion and the thrust washer.
11 For identification purposes, the following table indicates the number of teeth of each of the gearbox pinions.

Mainshaft pinions

Gear	**SB200**	**GT200X5**	**GT250X7**
1st (integral)	*12*	*12*	*12*
2nd	*17*	*16*	*16*
3rd	*21*	*20*	*19**
4th	*24*	*23*	*21**
5th	–	*24*	*22*
6th	–	–	*23*

**3rd and 4th gear pinions combined on X7 only*

Layshaft pinions

Gear	**SB200**	**GT200X5**	**GT250X7**
1st	*33*	*33*	*30*
2nd	*28*	*28*	*26*
3rd	*24*	*25*	*23*
4th	*21*	*23*	*21*
5th	–	*21*	*19*
6th	–	–	*18*

20.9a Fit mainshaft 5th gear pinion ...

20.9b ... and secure with thrust washer and circlip

20.9c Fit combined 3rd/4th gear pinion as shown

20.9d Position circlip and thrust washer on mainshaft

20.9e Slide the 6th gear pinion up against clip ...

20.9f ... then fit 2nd gear pinion and round wire circlip (arrowed)

20.9g Slide pinions outwards and fit circlip into groove

20.9h Bearing is preceded by thrustwasher and wave washers

20.10a Fit 2nd gear pinion to layshaft ...

20.10b ... and secure with thrust washer and circlip

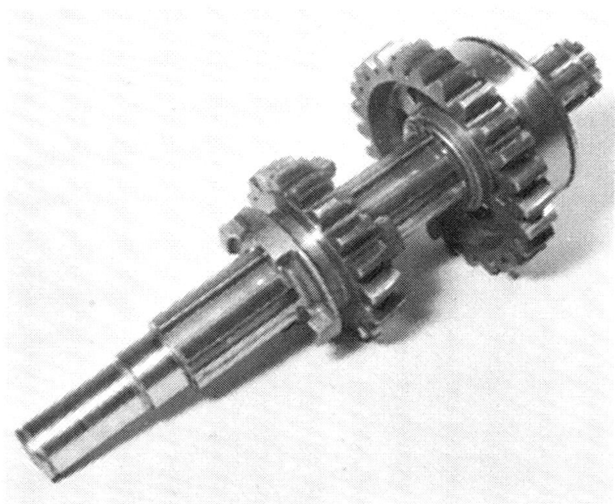
20.10c Fit the 6th gear pinion with selector groove outward

20.10d Secure with circlip, and fit thrust washer

20.10e Slide the 3rd gear pinion into place, ...

20.10f ... followed by the 4th gear pinion ...

20.10g ... which is secured by thrust washer and circlip

20.10h Fit the 5th gear pinion, selector groove inwards

20.10i Finish with the large 1st gear pinion, ...

20.10j ... thrustwasher and bearing

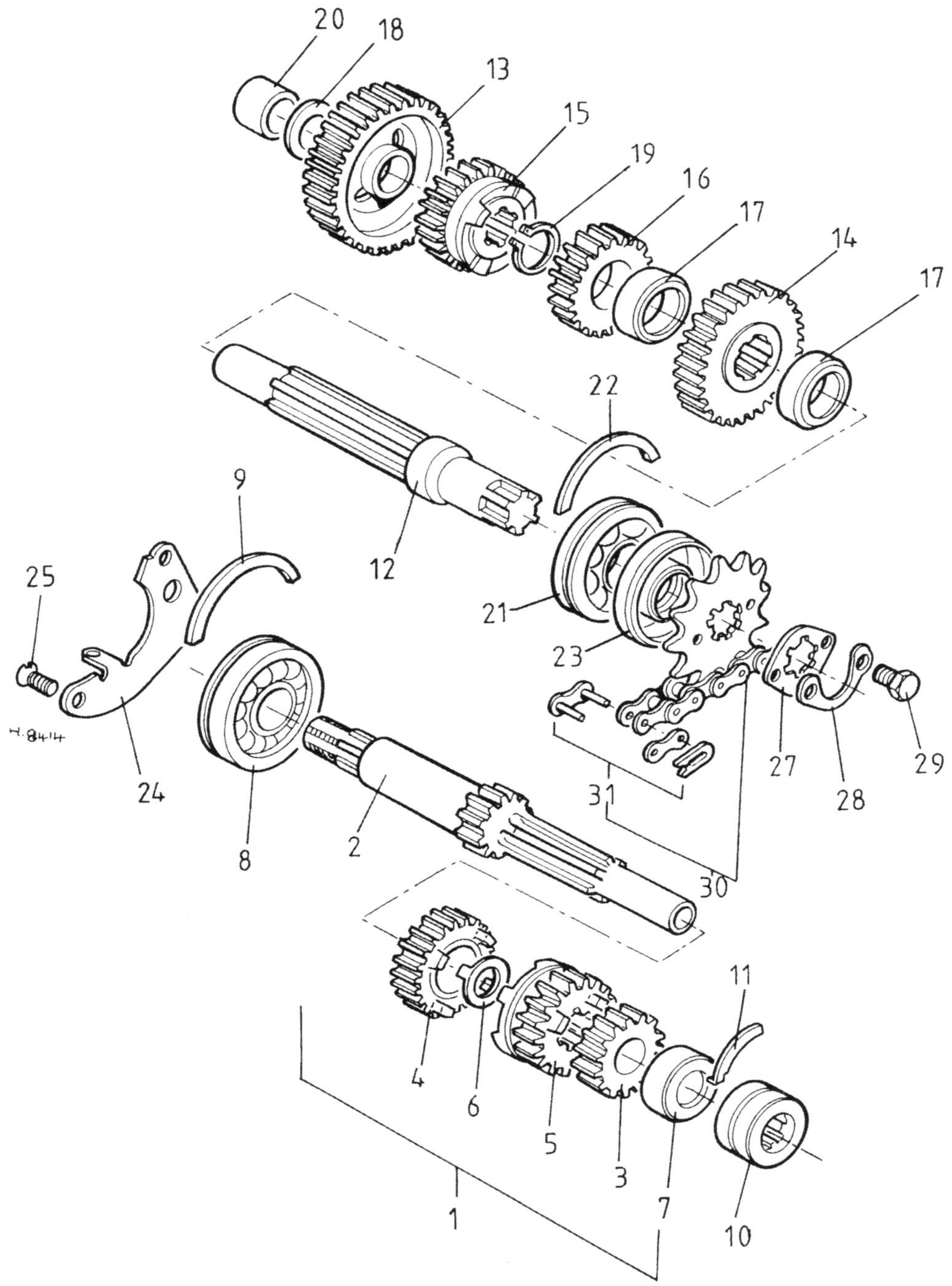

Fig. 1.7 Gearbox components – SB 200

1 *Mainshaft assembly*
2 *Mainshaft*
3 *Mainshaft 2nd gear pinion*
4 *Mainshaft 3rd gear pinion*
5 *Mainshaft 4th gear pinion*
6 *Circlip*
7 *Spacer*
8 *Right-hand bearing*
9 *Bearing locator clip*
10 *Left-hand bearing*
11 *Bearing locator clip*
12 *Layshaft*
13 *Layshaft 1st gear pinion*
14 *Layshaft 2nd gear pinion*
15 *Layshaft 3rd gear pinion*
16 *Layshaft 4th gear pinion*
17 *Distance collar – 2 off*
18 *Thrust washer*
19 *Circlip*
20 *Right-hand bush*
21 *Left-hand bearing*
22 *Bearing locator clip*
23 *Left-hand oil seal*
24 *Bearing retainer plate*
25 *Screw – 3 off*
26 *Final drive sprocket*
27 *Tab washer*
28 *Lock washer*
29 *Bolt – 2 off*
30 *Drive chain assembly*
31 *Chain link*

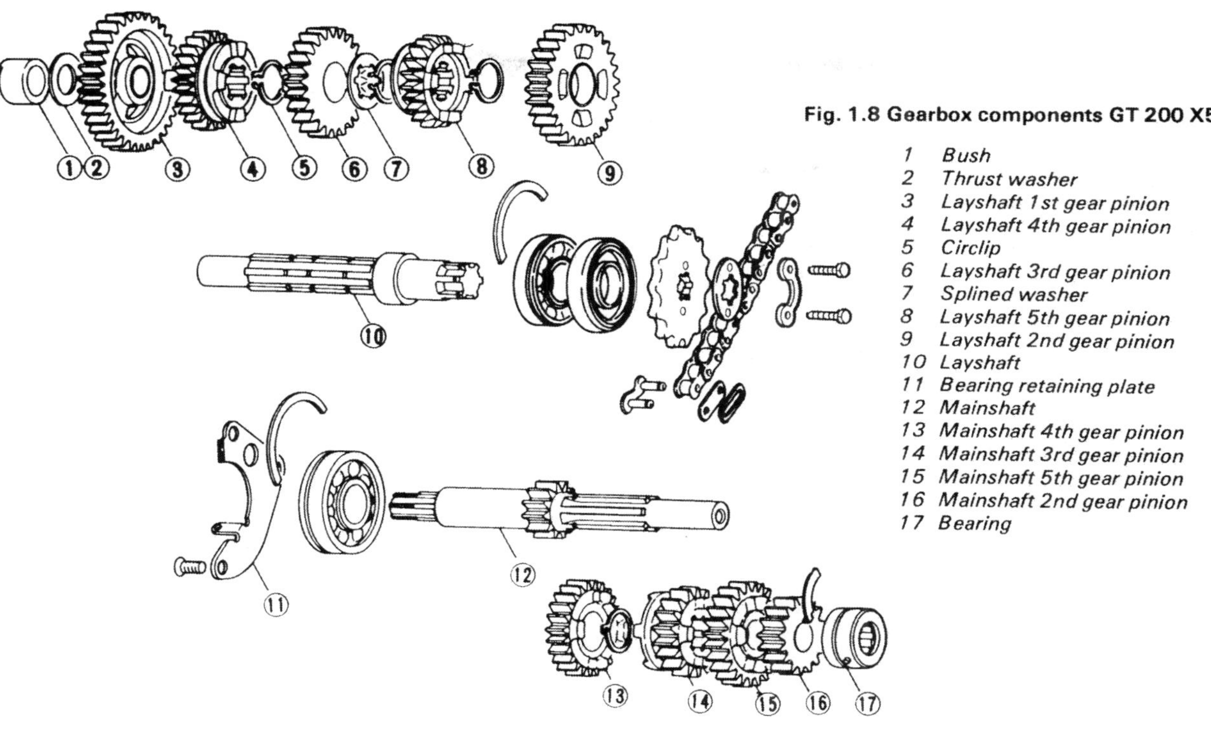

Fig. 1.8 Gearbox components GT 200 X5

1 *Bush*
2 *Thrust washer*
3 *Layshaft 1st gear pinion*
4 *Layshaft 4th gear pinion*
5 *Circlip*
6 *Layshaft 3rd gear pinion*
7 *Splined washer*
8 *Layshaft 5th gear pinion*
9 *Layshaft 2nd gear pinion*
10 *Layshaft*
11 *Bearing retaining plate*
12 *Mainshaft*
13 *Mainshaft 4th gear pinion*
14 *Mainshaft 3rd gear pinion*
15 *Mainshaft 5th gear pinion*
16 *Mainshaft 2nd gear pinion*
17 *Bearing*

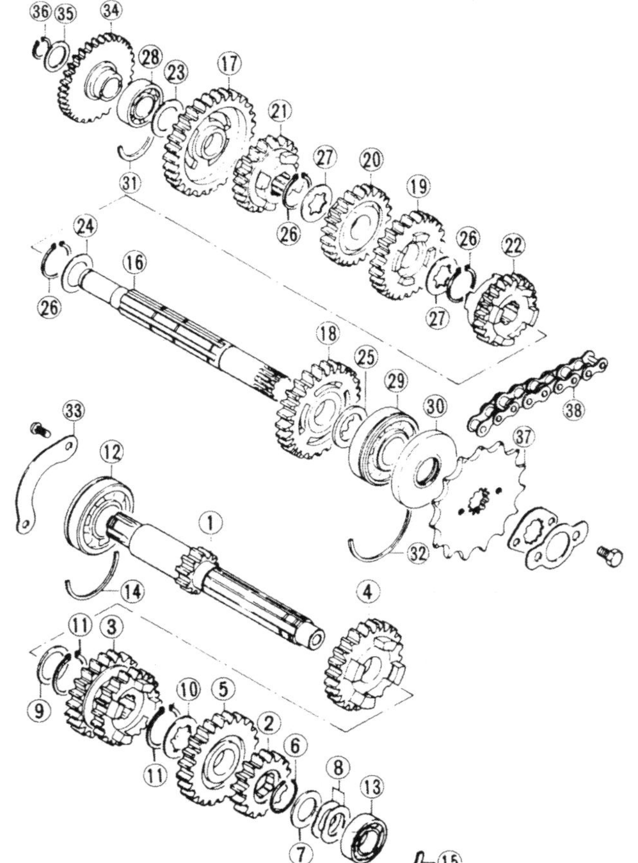

Fig. 1.9 Gearbox components – GT 250 X7

1 *Mainshaft*
2 *Mainshaft 2nd gear pinion*
3 *Mainshaft 3rd and 4th gear pinion*
4 *Mainshaft 5th gear pinion*
5 *Mainshaft 6th gear pinion*
6 *Circlip*
7 *Thrust washer*
8 *Wave washer – 2 off*
9 *Thrust washer*
10 *Splined washer*
11 *Circlip*
12 *Right-hand bearing*
13 *Left-hand bearing*
14 *Bearing half-ring*
15 *Bearing half-ring*
16 *Layshaft*
17 *Layshaft 1st gear pinion*
18 *Layshaft 2nd gear pinion*
19 *Layshaft 3rd gear pinion*
20 *Layshaft 4th gear pinion*
21 *Layshaft 5th gear pinion*
22 *Layshaft 6th gear pinion*
23 *Thrust washer*
24 *Thrust washer*
25 *Washer*
26 *Circlip*
27 *Splined washer*
28 *Right-hand bearing*
29 *Left-hand bearing*
30 *Left-hand oil seal*
31 *Bearing half-ring*
32 *Bearing half-ring*
33 *Mainshaft bearing retainer*
34 *Kick starter idler gear*
35 *Thrust washer*
36 *Circlip*
37 *Final drive sprocket*
38 *Final drive chain assembly*

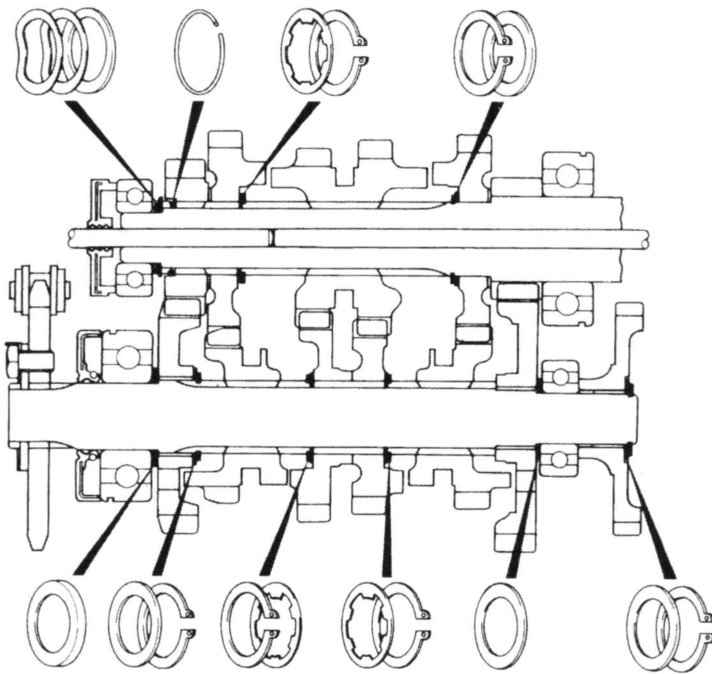

Fig. 1.10 Gearbox thrust washers and circlips – GT250 X7

21 Big-end and main bearings: examination and renovation

1 Failure of the big-end bearing is invariably accompanied by a knock within the crankcase that progressively becomes worse. Some vibration will also be experienced.

2 There should be no vertical play whatsoever in the big-end bearings, after the oil has been washed out. If even a small amount of vertical play is evident, the bearings are due for replacement. (A small amount of endfloat is both necessary and acceptable). Do not continue to run the machine with worn big-end bearings, for there is risk of breaking the connecting rods or crankshaft.

3 The built-up nature of the crankshaft assembly precludes the possibility of repair, as mentioned earlier. It will be necessary to obtain a replacement crankshaft assembly complete, under the Suzuki Service Exchange Scheme.

4 Failure of the main bearings is usually evident in the form of an audible rumble from the botton of the engine, accompanied by vibration that is felt through the footrests.

5 The crankshaft main bearings are of the ball journal type. If wear is evident in the form of play, or if the bearings feel rough as they are rotated, replacement is necessary. Always check after the old oil has been washed out of the bearings. Whilst it is possible to remove the outer bearings at each end of the crankshaft, it is probable that the centre bearing will also require attention. Here again it will be necessary to obtain a replacement crankshaft assembly, under the Suzuki Service Exchange Scheme.

6 Failure of both the big-end bearings and the main bearings may not necessarily occur as the result of high mileage covered. If the machine is used only infrequently, it is possible that condensation within the engine may cause premature bearing failure. The condition of the flywheels is usually the best guide. When condensation troubles have occurred, the flywheels will rust and become discoloured.

22 Oil seals: examination and renovation

1 The crankshaft oil seals form one of the most critical parts in any two-stroke engine because they perform the dual function of preventing oil from leaking along the crankshaft and preventing air from leaking into the crankcase when the incoming mixture is under crankcase vacuum during induction.

2 Oil seal failure is difficult to define precisely, although in most cases the machine will become difficult to start, particularly when warm. The engine will also tend to run unevenly and there will be a marked fall-off in performance, especially in the higher gears. This is caused by the intake of air into the crankcases which dilutes the mixture whilst it is in the crankcase, giving an exceptionally weak mixture for ignition.

3 It is possible to renew the outer crankcase oil seals without difficulty, but the inner seals form part of the crankshaft assembly. In this latter case, a replacement crankshaft assembly is the only way of ensuring a satisfactory crankcase seal.

4 It is unusual for the crankcase seals to become damaged during normal service, but instances have occurred when particles of broken piston rings have fallen into the crankcases and lacerated the seals. A defect of this nature will immediately be obvious.

21.5 Check the main and big end bearings for wear

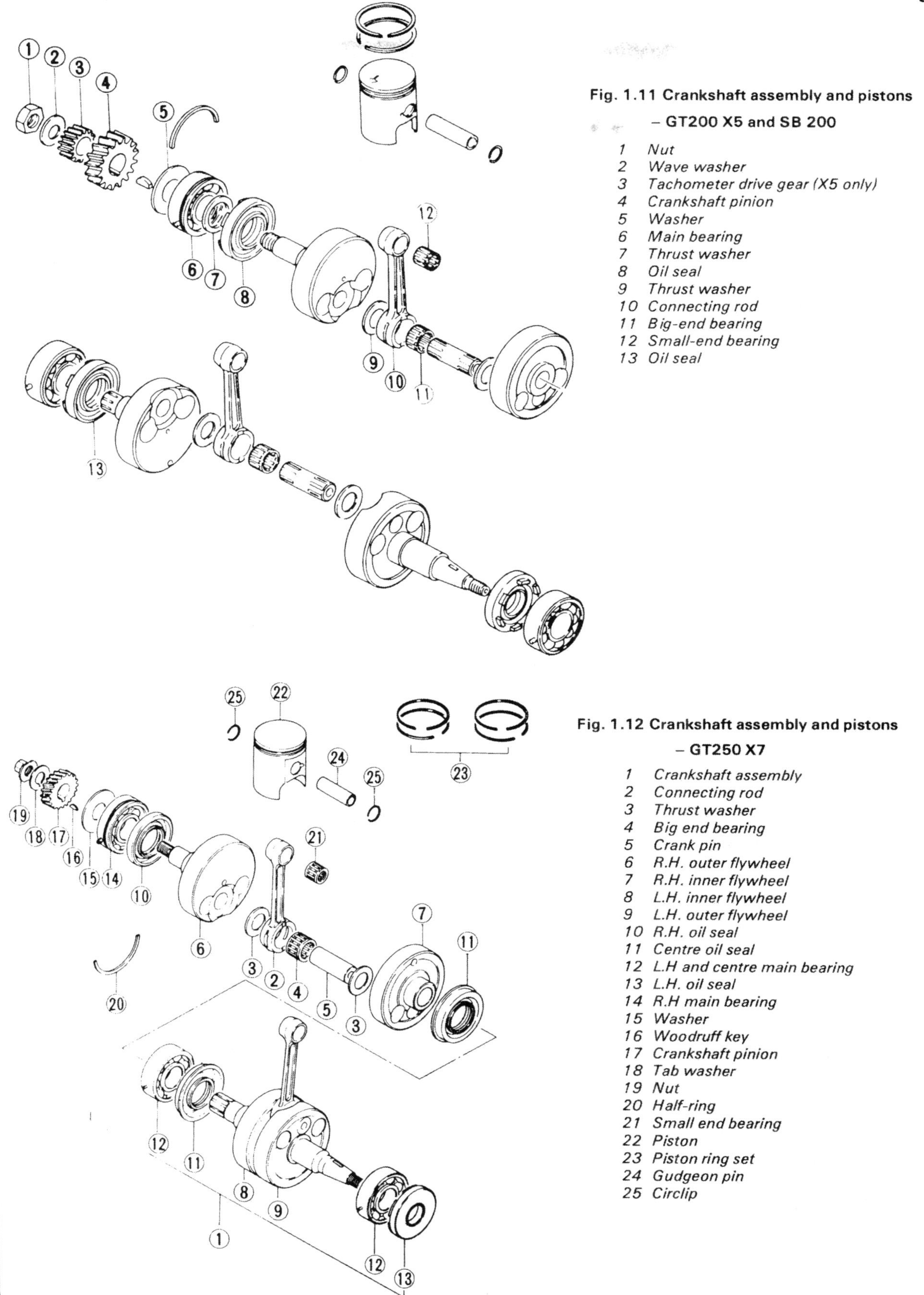

Fig. 1.11 Crankshaft assembly and pistons – GT200 X5 and SB 200

1 *Nut*
2 *Wave washer*
3 *Tachometer drive gear (X5 only)*
4 *Crankshaft pinion*
5 *Washer*
6 *Main bearing*
7 *Thrust washer*
8 *Oil seal*
9 *Thrust washer*
10 *Connecting rod*
11 *Big-end bearing*
12 *Small-end bearing*
13 *Oil seal*

Fig. 1.12 Crankshaft assembly and pistons – GT250 X7

1 *Crankshaft assembly*
2 *Connecting rod*
3 *Thrust washer*
4 *Big end bearing*
5 *Crank pin*
6 *R.H. outer flywheel*
7 *R.H. inner flywheel*
8 *L.H. inner flywheel*
9 *L.H. outer flywheel*
10 *R.H. oil seal*
11 *Centre oil seal*
12 *L.H and centre main bearing*
13 *L.H. oil seal*
14 *R.H main bearing*
15 *Washer*
16 *Woodruff key*
17 *Crankshaft pinion*
18 *Tab washer*
19 *Nut*
20 *Half-ring*
21 *Small end bearing*
22 *Piston*
23 *Piston ring set*
24 *Gudgeon pin*
25 *Circlip*

23 Cylinder barrels: examination and renovation

1 The usual indication of badly worn cylinder barrels and pistons is piston slap, a metallic rattle that occurs when there is little or no load on the engine. If the top of the bore of the cylinder barrels is examined carefully, it will be found that there is a ridge on the thrust side, the depth of which will vary according to the amount of wear that has taken place. This marks the limit of travel of the uppermost piston ring.
2 Measure the bore diameter just below the ridge, using an internal micrometer. Compare this reading with the diameter at the bottom of the cylinder bore, which has not been subjected to wear. If the difference in readings exceeds 0.05 mm (0.002 inch) the cylinder should be rebored and fitted with an oversize piston and rings.
3 Suzuki can provide pistons in two oversizes: 0.5 mm (0.020 in), and 1.0 mm (0.040 in). The 1.0 mm oversize is the limit to which the cylinder can be overbored with safety.
4 Check that the surface of the cylinder bores is free from score marks or other damage that may have resulted from an earlier engine seizure or a displaced gudgeon pin. A rebore will be necessary to remove any deep indentations, irrespective of the amount of bore wear that has taken place, otherwise a compression leak will occur.
5 Make sure the external cooling fins of the cylinder barrels are not clogged with oil or road dirt, which will otherwise prevent the free flow of air and cause the engine to overheat. Remove any carbon that has accumulated in the exhaust ports, using a blunt-ended scraper so that the surface of the ports is not scratched. Finish off with metal polish so that the ports have a smooth, shiny appearance. This will aid gas flow and prevent carbon from adhering so firmly on future occasions.
6 Under no circumstances modify or re-profile the ports in the search for extra performance. The size and location of the ports is critical in terms of engine performance and the dimensions chosen have been selected to give good performance consistent with a high standard of mechanical reliability.
7 If the cylinder barrels have been rebored, it will be necessary to round off the extreme edges of the ports, to prevent rapid wear of the piston rings. Use a scraper tool or hand grinder and finish off with fine emery cloth.

24 Cylinder barrels: reed valve assembly – X7 model

1 A reed valve assembly is mounted to the underside of each cylinder barrel, forming a supplementary induction system, in addition to the normal piston porting arrangement. The normal inlet tract is designed to open later and close earlier than is normal in engines of this type. This enhances low to intermediate engine speed performance, at the expense of high-speed efficiency. The port timing is assisted by the reed valve, however, as this opens and closes independently of the piston position, and according to the demands of the engine. In practice, this means that the reed valve applies a boost to top end power, giving a wider spread of performance than a conventional piston-ported unit.
2 The reed valve is automatic in operation and requires no maintenance. Care must be taken not to damage the assembly, as it is delicate, and can easily be rendered inoperative if it is allowed to become dirty. For this reason, remove the two units if the cylinder barrel is being worked on, to avoid any possible damage.
3 Each reed valve unit is secured by two cross-head screws, and is fitted with a gasket which must be renewed if the valve has been disturbed. Take great care not to drop the valve unit, as the thin petals are easily damaged. When refitting the valves, a thread locking fluid should be used on the retaining screws.

25 Pistons and piston rings: examination and renovation

1 If a rebore is necessary, the existing pistons and piston rings can be disregarded because they will have to be replaced with their new oversize equivalents as a matter of course.
2 Remove all traces of carbon from the piston crowns, using a blunt-ended scraper to avoid scratching the surface. Finish off by polishing the crowns with metal polish, so that carbon will not adhere so readily in the future. Never use emery cloth on the soft aluminium.
3 Piston wear usually occurs at the skirt or lower end of the piston and takes the form of vertical streaks or score marks on the thrust face. There may also be some variation in the thickness of the skirt, in an extreme case.
4 The piston ring grooves may have become enlarged in use, allowing the rings to have greater side float. If the clearance exceeds 0.006 inch, the pistons are due for replacement. It is unusual for this amount of wear to occur on its own.
5 Piston ring wear is measured by removing the rings from the piston and inserting them in the cylinder, using the crown of a piston to locate them about $1\frac{1}{2}$ inches from the top of the bore. Make sure they rest squarely in the bore. Measure the end gap with a feeler gauge; if the gap exceeds that given in the Specifications, the rings must be replaced.
6 The pistons are fitted with Keystone rings, which can be identified by their shape. These rings are not interchangeable with the conventional type of rings and must be used only in conjunction with Keystone pistons. They have a 7° taper on their uppermost surface. They are handled in the same fashion as conventional rings.

24.1a Reed valve units are fitted to each barrel on X7 model

24.1b Valve is retained by two cross-headed screws

25.2 Piston crowns should be cleaned off. Note arrow

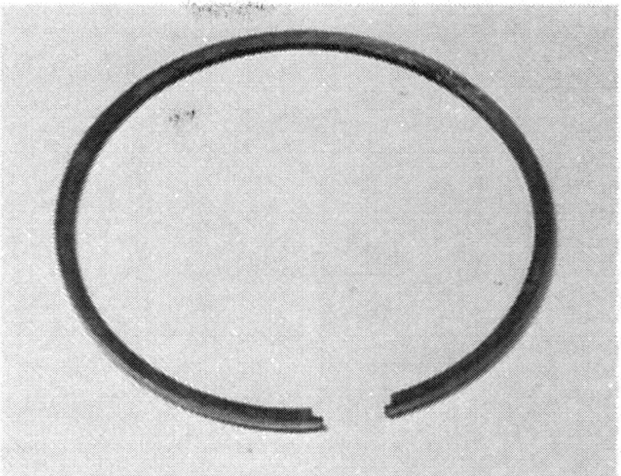

25.5 Check rings for wear. Note locating cut-outs

26 Cylinder head: examination and renovation

1 Remove all traces of carbon from the cylinder head, using a blunt-ended scraper. Finish by polishing with metal polish, to give a smooth, shiny surface. This will aid gas flow and will also prevent carbon from adhering so firmly in the future.
2 Check the condition of the threads in the sparking plug holes. If the threads are worn or stretched as the result of over-tightening the plugs, they can be reclaimed by a "Helicoil" thread insert. Most dealers have the means of providing this cheap but effective repair.
3 Make sure the cylinder head fins are not clogged with oil or road dirt, otherwise the engine will overheat. If necessary, use a wire brush to clean the cooling fins.
4 Lay the cylinder head on a sheet of plate glass to check for distortion. Aluminium alloy cylinder heads will distort very easily, especially if the cylinder head bolts are tightened down unevenly. If the amount of distortion is only slight, it is permissible to rub the head down until it is flat once again by wrapping a sheet of very fine emery cloth around the plate glass sheet and rubbing with a rotary motion.
5 If the cylinder head is distorted badly, it is advisable to fit a new replacement. Although the head joint can be restored by skimming, this will raise the compression ratio of the engine and may adversely affect performance.

27 Gearbox components: examination and renovation

1 Give the gearbox components a close visual inspection for signs of wear or damage such as broken or chipped teeth, worn dogs, damaged or worn splines and bent selectors. Replace any parts found unserviceable because they cannot be reclaimed in a satisfactory manner.
2 Check the condition of the various pawl springs and the kickstart return spring. If any of the pawl springs fail after the engine has been rebuilt, a further complete stripdown of the engine unit will be necessary. In the case of GT250 X7 models this also applies to the kickstart spring.
3 Check the condition of the internal serrations in the kickstart pinion. If these serrations become worn or the edges rounded, the kickstart will slip. Replacement of the pinion is the only means of restoring the full action.
4 Check also that the tip of the kickstart pawl is not worn because this too will promote slip during engagement. If the kickstart pinion is renewed, it is good policy to renew also the pawl.
5 Make sure that the selector stopper roller is in good shape and is free from any 'flats'. If in doubt, renew.
6 The gearbox bearings must be free from play and show no signs of roughness when they are rotated. Each shaft has a ball journal bearing at one end and a caged needle roller bearing at the other.
7 It is advisable to renew the gearbox oil seals irrespective of their condition. Should a re-used oil seal fail at a later date, a considerable amount of dismantling is necessary to gain access and renew it.
8 Check the gear selector rods for straightness by rolling them on a sheet of plate glass. A bent rod will cause difficulty in selecting gears and will make the gear change action particularly heavy.
9 The selector forks should be examined closely, to ensure that they are not bent or badly worn. Wear is unlikely to occur unless the gearbox has been run for a period with a particularly low oil content.
10 The tracks in the gear selector drum, with which the selector forks engage, should not show any undue signs of wear unless neglect has led to under lubrication of the gearbox. Check that the plunger spring bearing on the cam plate plunger has not lost its action and that the springs of the gear change lever pawl assembly have good tension. Any damage to, or weakness of, the gear change lever return spring will be self-evident.

27.1 Examine gear teeth and selector grooves for wear

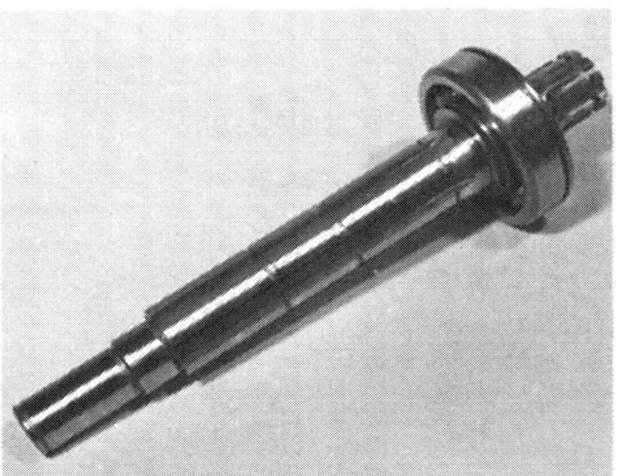

27.6a Wash bearings in petrol and spin to check for wear

27.6b Outer races have locating grooves

28 Clutch assembly: examination and renovation

1 After an extended period of service, the friction plates will have become worn sufficiently to warrant renewal, to avoid subsequent problems with clutch slip. The lining thickness is measured across the friction plate using a vernier caliper. When new, each plate measures 2.9–3.1 mm (0.11–0.12 in) on 200cc models and 3.4–3.6 mm (0.134–0.141 in) on 250cc models. If the friction plates are worn by more than 0.3 mm (0.012 in), the plates should be renewed as a complete set.

2 The plain plates should be free from any signs of blueing, which would indicate that the clutch had overheated in the past. Check each plate for distortion by laying it on a flat surface, such as a sheet of plate glass or similar, and measuring any detectable gap using feeler gauges. The plates must be less than 0.1 mm (0.004 in) out of true.

3 The clutch springs may, after a considerable mileage, require renewal, and their free length should be checked as a precautionary measure. Those used on the X7 are nominally 36.9–38.4 mm (1.45–1.51 in) in length, and should be renewed if compressed to 36.5 mm (1.44 in) or less. The tension springs used on the X5 and SB200 are 32.0 mm (1.26 in) long when new, and will require renewal if they have stretched to 33.6 mm (1.32 in) or more.

4 Check the caged needle roller thrust race for signs of wear or damage. This bearing can normally be expected to give reliable service for many miles. As it is crucial to the smooth operation of the clutch, check carefully for signs of damage or cracking of the race or rollers, renewing the bearing if any defect is discovered.

5 Check the condition of the slots in the outer surface of the clutch centre and the inner surfaces of the outer drum. In an extreme case, clutch chatter may have caused the tongues of the inserted plates to make indentations in the slots of the outer drum, or the tongues of the plain plates to indent the slots of the clutch centre. These indentations will trap the clutch plates as they are freed and impair clutch action. If the damage is only slight the indentations can be removed by careful work with a file and the burrs removed from the tongues of the clutch plates in similar fashion. More extensive damage will necessitate renewal of the parts concerned.

6 The clutch release mechanism attached to the inside of the left-hand crankcase cover does not normally require attention, provided it is greased at regular intervals. It is held to the cover by two cross-head screws and operates on the worm and quick start thread principle. A light return spring ensures that the pressure is taken from the end of the clutch push rod when the handlebar lever is released and the clutch fully engaged.

28.6 Clutch release mechanism rarely causes problems

29 Engine reassembly: general

1 Before reassembly of the engine/gear unit is commenced, the various component parts should be cleaned thoroughly and placed on a sheet of clean paper, close to the working area.

2 Make sure all traces of old gaskets have been removed and that the mating surfaces are clean and undamaged. One of the best ways to remove old gasket cement is to apply a rag soaked in methylated spirit. This acts as a solvent and will ensure that the cement is removed without resort to scraping and the consequent risk of damage.

3 Gather together all the necessary tools and have available an oil can filled with clean engine oil. Make sure all new gaskets and oil seals are to hand, also all replacement parts required. Nothing is more frustrating than having to stop in the middle of a reassembly sequence because a vital gasket or replacement has been overlooked.

4 Make sure that the reassembly area is clean and that there is adequate working space. Refer to the torque and clearance settings wherever they are given. Many of the smaller bolts are easily sheared if over-tightened. Always use the correct sized screwdriver bit for the cross-head screws and never an ordinary screwdriver or punch. If the existing screws show evidence of

maltreatment in the past, it is advisable to renew them as a complete set.

5 In addition to any replacement parts, gasket cement etc, it will be necessary to obtain a bottle or tube of thread locking compound such as Loctite. This substance is used widely throughout the engine unit.

30 Engine reassembly: refitting the selector drum and forks – X5 and SB200

1 Place the upper crankcase half on the workbench, with the crankcase mouths downwards. Lubricate the selector drum, and slide it into position in the casing. It will be noted that there is a depression machined into the face of the drum, and this should be positioned to coincide with the threaded hole for the detent plunger. Fit the detent assembly, noting that the correct sealing washer must be used to ensure that the detent plunger is accurately positioned.

2 When refitting the selector forks, the following points should be noted; on all models, neutral should be selected by positioning the neutral detent plunger in engagement with its recess in the selector drum. On SB200 machines, two different selector forks are employed. The fork fitted nearest the front of the engine unit is numbered 25221–10200, whilst the rear-most item is numbered 25211–10200. The appropriate line drawing shows the forks in relation to the selector drum, note however, that this shows the arrangement in its normal working position, whereas reassembly is carried out with the casing inverted.

3 The X5 model, having a five-speed gearbox, requires three selector forks. Two of these are identical, and share the rear-most support rod. The remaining fork is fitted to the front of the drum on the other support rod.

4 Lubricate the support rods and slide them into position, arranging the selector forks as the rod is installed. Ensure that the locating pins on the forks engage in the selector drum tracks, and lubricate the latter before moving on to the reassembly of the remaining gearbox components.

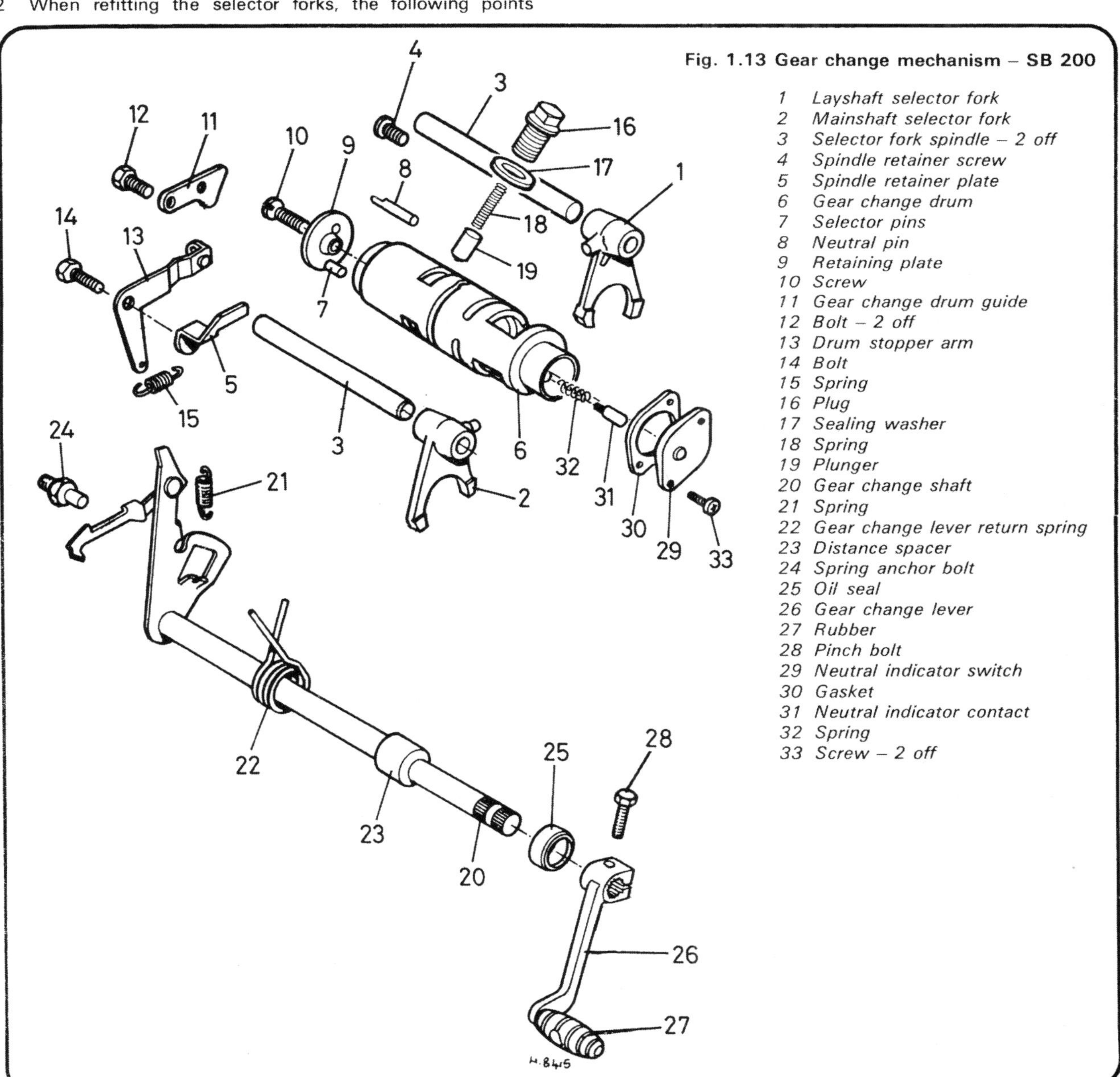

Fig. 1.13 Gear change mechanism – SB 200

1 *Layshaft selector fork*
2 *Mainshaft selector fork*
3 *Selector fork spindle – 2 off*
4 *Spindle retainer screw*
5 *Spindle retainer plate*
6 *Gear change drum*
7 *Selector pins*
8 *Neutral pin*
9 *Retaining plate*
10 *Screw*
11 *Gear change drum guide*
12 *Bolt – 2 off*
13 *Drum stopper arm*
14 *Bolt*
15 *Spring*
16 *Plug*
17 *Sealing washer*
18 *Spring*
19 *Plunger*
20 *Gear change shaft*
21 *Spring*
22 *Gear change lever return spring*
23 *Distance spacer*
24 *Spring anchor bolt*
25 *Oil seal*
26 *Gear change lever*
27 *Rubber*
28 *Pinch bolt*
29 *Neutral indicator switch*
30 *Gasket*
31 *Neutral indicator contact*
32 *Spring*
33 *Screw – 2 off*

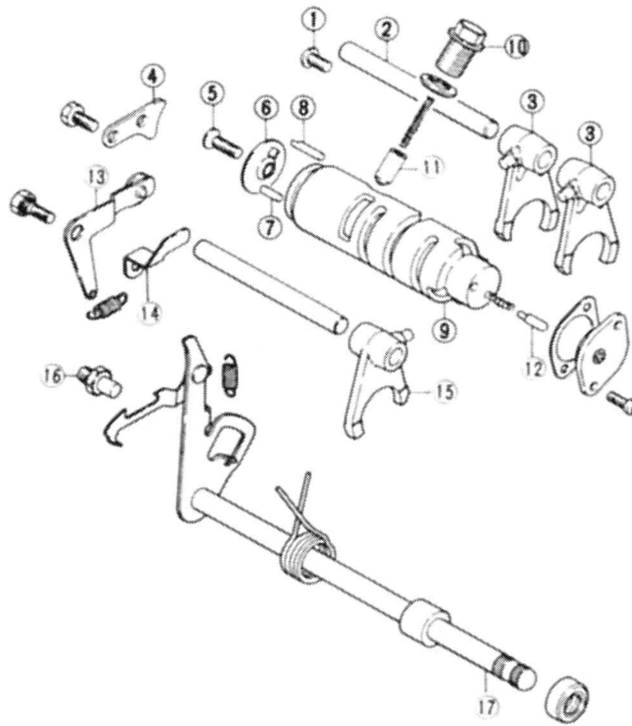

Fig. 1.14 Gear change mechanism – GT200 X5

1 *Screw*
2 *Selector fork shaft*
3 *Selector fork*
4 *Guide plate*
5 *Screw*
6 *Pin retainer*
7 *Selector pins*
8 *Neutral stopper pin*
9 *Selector drum*
10 *Housing*
11 *Plunger*
12 *Neutral switch pin*
13 *Stopper arm*
14 *Retainer plate*
15 *Selector fork*
16 *Special locating bolt*
17 *Selector shaft assembly*

31 Engine reassembly: refitting the selector drum and forks – X7 model

1 If the selector drum has been dismantled for inspection, fit the small thrust washer, cam plate and retaining circlip to the left-hand end of the drum, noting that the cam plate projections must face inwards, and that the plate locates on a small dowel pin which projects from the drum surface.

2 Arrange the lower crankcase half on the workbench, lubricate the selector drum and its needle roller bearing, and slide the drum into position. It will be noted from the accompanying line drawing, that all three of the selector forks differ in design, and care must be taken to ensure that they are installed as shown. The forks can be identified as follows: the 3rd and 4th gear fork is fitted to the front support rod. It has light webs around its boss, and the locating peg is well offset in relation to the fork.

3 The 5th gear fork shares the rearmost support rod with the 6th gear fork, both forks operating on the layshaft cluster. The 5th gear fork's locating peg is almost in line with the fork, which is heavily webbed around the boss. This fork is fitted to the **right** of the support rod. The remaining, 6th gear, fork is easily identified by the cranked shape of the fork. The locating peg is offset.

4 It is important that the single 3rd and 4th gear fork is fitted first. Slide the rod into position, installing the fork as it is pushed through. Before the rod enters its blind bore, position the stopper arm assembly with the widest portion of its boss facing the fork. The stopper arm return spring can be grasped with the aid of pointed-nose pliers, and hooked over its anchoring pin beneath the selector drum.

5 Slide the rear support rod into position, fitting the 5th gear fork first, followed by the 6th gear fork. Check that the forks are engaged in their selector drum grooves, and lubricate the latter thoroughly before proceeding further.

31.1a Cam plate is fitted as shown. Note locating pin

31.1b Cam plate is secured by a circlip

31.2a Place stopper arm assembly in casing as shown ...

31.2b ... then slide the selector drum into position

31.4a Slide support rail through selector fork and stopper arm

31.4b Use pointed-nosed pliers to attach spring as shown

31.5a Fit rear selector forks and support rod

31.5b The selector forks should be arranged as shown. Note the bearing location half-rings (arrowed)

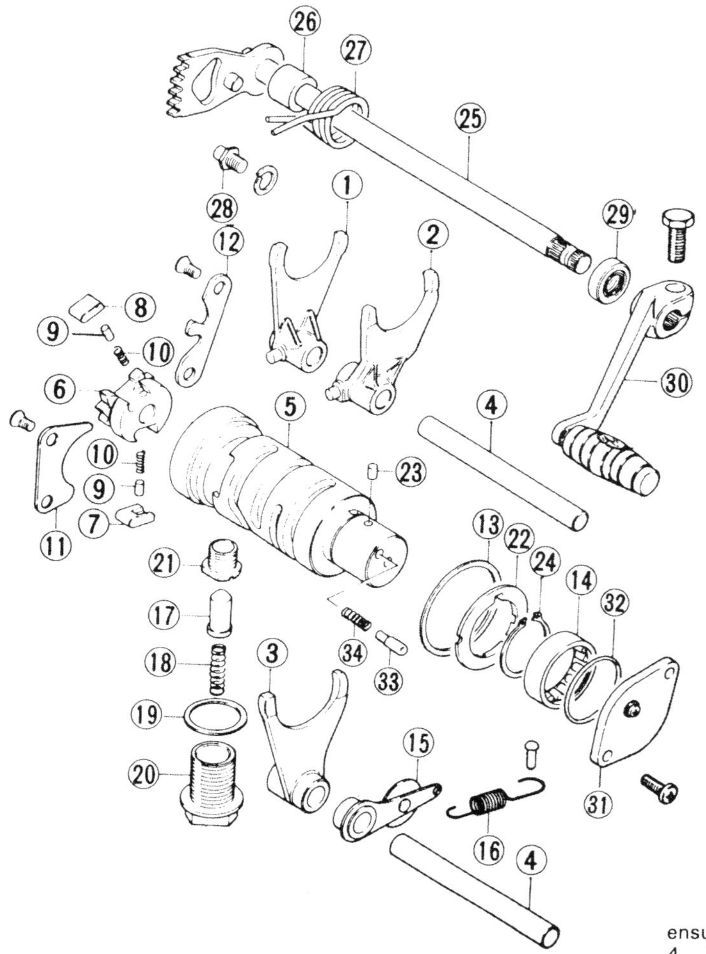

Fig. 1.15 Gear change mechanism – GT 250 X7

1 *Selector fork (layshaft 5th gear)*
2 *Selector fork (layshaft 6th gear)*
3 *Selector fork (mainshaft 3rd/4th gear)*
4 *Selector fork shaft*
5 *Selector drum*
6 *Ratchet assembly*
7 *Pawl*
8 *Pawl*
9 *Pin*
10 *Spring*
11 *Pawl plate*
12 *Guide*
13 *Thrust washer*
14 *Bearing*
15 *Stopper arm*
16 *Spring*
17 *Detent plunger*
18 *Spring*
19 *Sealing washer*
20 *Drain plug*
21 *Guide*
22 *Cam plate*
23 *Locating pin*
24 *Circlip*
25 *Gearchange shaft*
26 *Spacer*
27 *Centring spring*
28 *Locating bolt*
29 *Oil seal*
30 *Gearchange lever*
31 *Neutral switch*
32 *Sealing ring*
33 *Pin*
34 *Spring*

32 Engine reassembly: refitting the gearbox shafts, kickstart mechanism and crankshaft

1 Fit the half-ring retainers into their appropriate slots in the casing. It is important that these are not inadvertently omitted, as they provide axial location of the various shafts. Place the gearbox mainshaft and layshaft assemblies in position, ensuring that the selector forks engage in the grooves in the pinions. Lubricate the gear teeth and shafts, and check that the gear pinions rotate smoothly. On SB200 and X5 models, it will be noted that the layshaft bush, on the right-hand side of the shaft, has a stepped end. This must be positioned carefully so that it locates against the mainshaft bearing retainer after the crankcase halves are joined. To this end, the step should be at 90° to the gasket face with the lower portion facing the mainshaft.

2 On SB200 and X5 machines, the kickstart mechanism is assembled and installed as follows. Fit the circlip to the splined end of the kickstart shaft, arranging it so that the ends of the circlip are supported by the raised splines and do not fall either side of the spline. Fit the thrust washer and the light compression spring over the splined end of the shaft.

3 Examination of the end of the shaft will reveal an alignment dot between two of the raised splines. A similar dot will be found on the boss of the kickstart ratchet, and when fitting the latter, the two marks must coincide. Slide the kickstart pinion into position, followed by the headed support bush, head inward. From the opposite end of the shaft, fit the plain bush, ensuring that the small locating pin is in position.

4 Ensure that the ratchet guide plate is secured to the inside of the crankcase. If it has been removed, reassemble it using a thread locking compound on the screws to prevent their loosening during use. Place the kickstart shaft assembly into the casing, ensuring that the locating peg in the right-hand support bush engages in its hole in the casing. Check that when the kickstart shaft is rotated to its normal resting position, the ratchet is pulled clear of the pinion by the ratchet guide plate.

5 In the case of the X7 model, assemble the thrust washer, spring retainer and return spring on the shaft end. Engage the internal tang of the spring in its hole in the shaft, then fit the plastic spring guide and secure it with its circlip. Place the assembly in its recess in the casing, ensuring that the large thrust washer is fitted to the outside of the crankcase. The oil pump drive pinion should be fitted at this stage. It will be noted that whilst most models employ an externally bushed pinion, some early models may be found to employ a slightly different arrangement. The two types are not interchangeable.

6 When placing the crankshaft in position, the following points should be noted. Where new oil seals are to be fitted, ensure that the spacing tangs on the seals face toward their respective bearings. The seal lips should be greased prior to installation. Locking fluid, Suzuki part number 99000-32040 or equivalent, should be applied to the outer face of each seal to prevent movement in service, and to ensure a gas-tight joint. Note that each of the three main bearings has a locating pin which must be arranged to fit into the appropriate recess in the casing.

7 On later SB200 and X5 models do not forget to fit a new O-ring against the layshaft left-hand bearing inner race, followed by the gearbox sprocket spacer. Take care not to damage the oil seal lips as the spacer is refitted.

32.1a Fit mainshaft assembly, ensuring that selector fork engages

32.1b The completed gearbox assembly should look like this

32.1c Fit new oil seals using locking fluid

32.5a Fit the washer, spring guide and spring ...

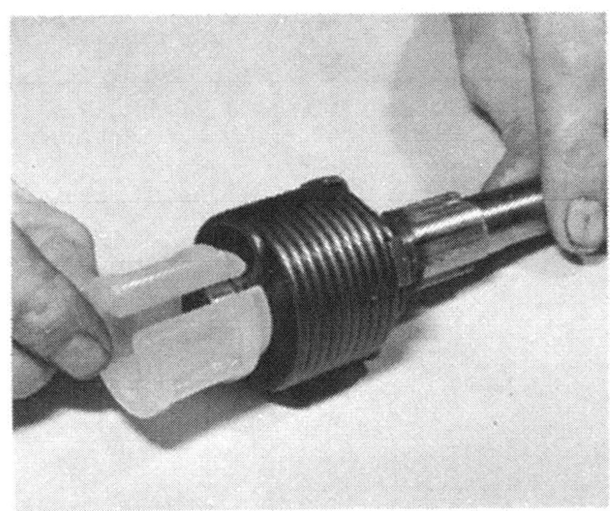
32.5b ... followed by the plastic spring guide and circlip

32.5c Assembly is fitted as shown – note thrustwasher

32.5d Oil pump drive pinion and bush fits thus

32.6a Lower the crankshaft assembly into position

32.6b Bearings have small roll pins ...

32.6c ... which locate bearings in casing

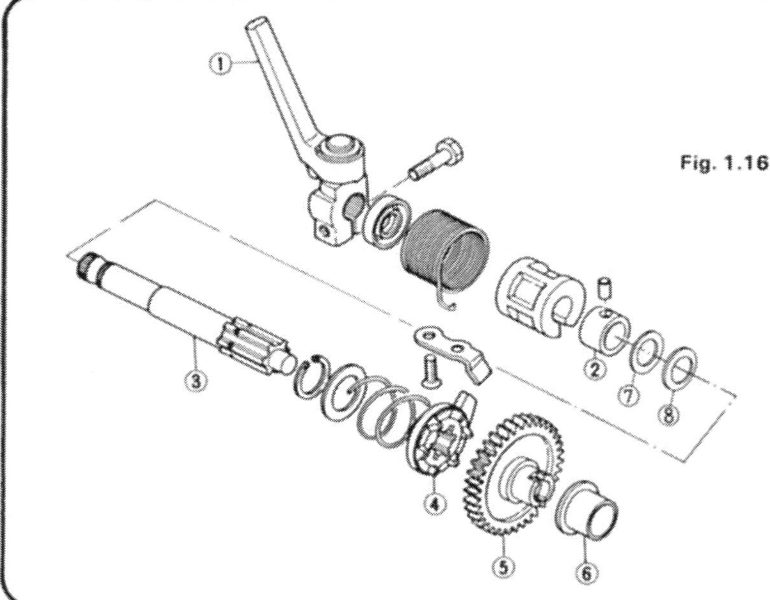

Fig. 1.16 Kickstart mechanism – GT200 X5 and SB200

1 *Kickstart lever*
2 *Bush*
3 *Kickstart shaft*
4 *Ratchet*
5 *Kickstart pinion*
6 *Bush*
7 *Thrust washer (SB200 X only)*
8 *Wave washer (SB200 X only)*

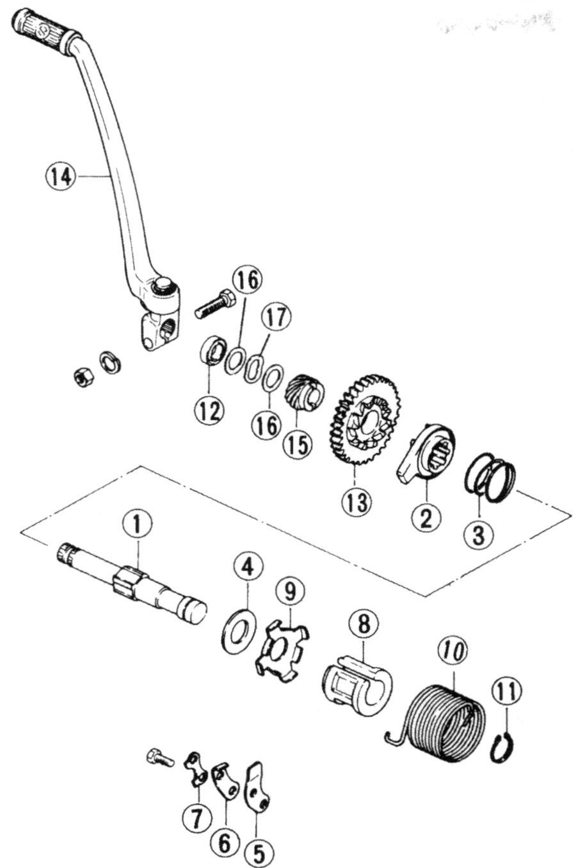

Fig. 1.17 Kickstart mechanism – GT250 X7

1	*Kickstart shaft*	*9*	*Spring seat*
2	*Kickstart ratchet*	*10*	*Return spring*
3	*Spring*	*11*	*Circlip*
4	*Washer*	*12*	*Oil seal*
5	*Guide plate*	*13*	*Kickstart pinion*
6	*Stop*	*14*	*Kickstart lever*
7	*Tab washer*	*15*	*Tachometer drive pinion*
8	*Spring guide*	*16*	*Thrust washer*
		17	*Wave washer*

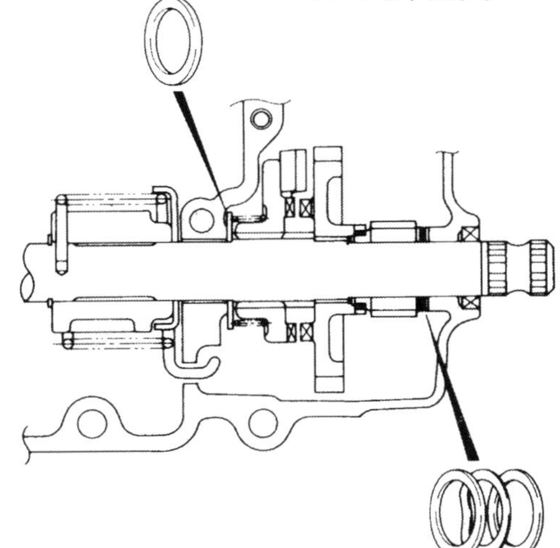

Fig. 1.18 Thrust washer arrangement for kickstart mechanism – GT 250 X7

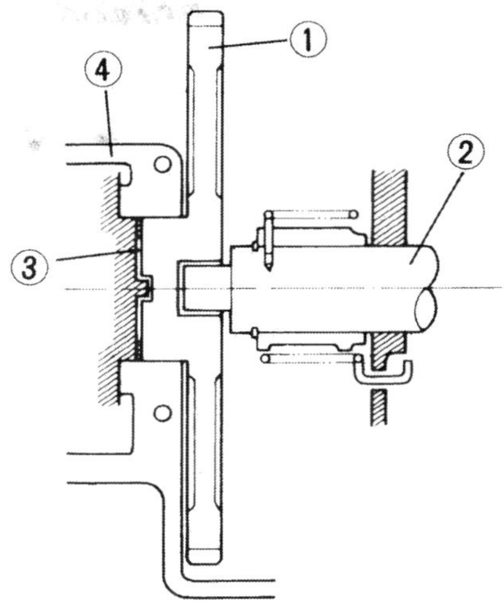

Fig. 1.19 Oil pump pinion – GT250 X7 up to engine number 108072

1 Pinion	*3 Thrust washer*
2 Kickstart shaft	*4 Crankcase*

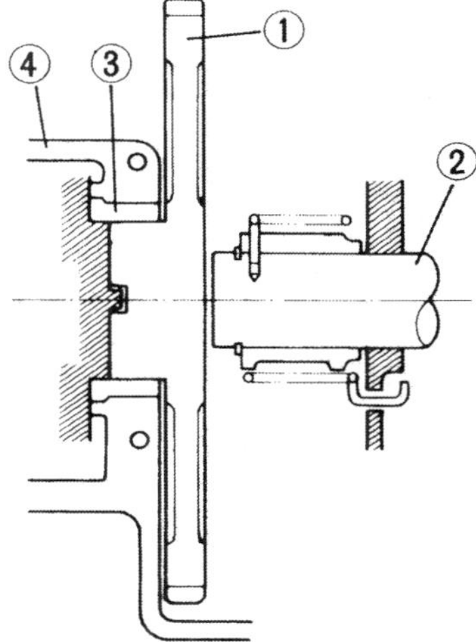

Fig. 1.20 Oil pump pinion – GT250 X7 after engine number 108072

1 Pinion	*3 Bush*
2 Kick start shaft	*4 Crankcase*

33 Engine reassembly: joining the crankcase halves

1 Check that the various moving parts are adequately lubricated with engine oil. The joint faces of each crankcase half should be meticulously cleaned, preferably with methylated spirit, to remove any residual oil or moisture. Coat one mating surface with an even film of silicon RTV jointing compound, Hermetite Instant Gasket, Suzuki Bond No 4 or similar. The compound should be left for the amount of time prescribed by the manufacturer, usually about ten minutes, before assembling the crankcase.

2 Check that the locating dowels are in position, then offer up the empty casing half. On X7 models, guide the connecting rods through the crankcase mouths. Fit the various crankcase bolts fingertight, then commence tightening in the sequence denoted by the numbers cast into the casing. Each bolt should be tightened a little at a time to prevent warping, going through the

sequence until all the bolts are fully tightened. The correct torque settings are as follows:

Torque settings for crankcase bolts

Bolt size	**GT250 X7**	**SB200, X5**
6mm	*0.6-0.9 kgf m (4.5-6.5 lbf ft)*	*0.6-1.0 kgf m (4.5-7.5 lbf ft)*
8mm	*1.5-2.0 kgf m (11.0-14.5 lbf ft)*	*1.8-2.3 kgf m (13.0-16.5 lbf ft)*

All bolts to be tightened in sequence denoted by crankcase numbers. On X7 models, do not omit wiring clips on upper crankcase.

3 Before proceeding further, check that each of the gearbox shafts and the crankshaft are free to turn. If this is not the case, separate the cases and rectify the problem before moving on.

34 Engine reassembly: installing the gear selector mechanism – SB200 and X5

1 Fit the gearbox mainshaft retaining plate, noting that the plate also acts as a means of location for the layshaft bush, which has a corresponding step machined into its outer face (see Section 32.1). The three securing screws should be tightened after applying locking compound to their threads. Turn the selector drum, if necessary, to position the indentation in the selector drum immediately beneath the threaded hole for the neutral detent plunger. Fit and tighten the detent assembly ensuring that the correct sealing washer is used. This is important, as the wrong thickness of washer could result in the selector drum becoming locked or the detent mechanism being rendered ineffective.

2 Fit the pins in the end of the selector drum, noting that the neutral pin is shaped differently from the others. Refit the end plate, and tighten the securing screw, having first applied some locking fluid to the screw threads. Reassemble the selector drum locating plate, again using thread locking fluid on its two retaining bolts.

3 Before fitting the gearchange shaft and claw assembly, check that the centring spring is arranged correctly in relation to the tang. The two ends of the spring should be parallel, and must not cross over. If not yet in position, fit the screw which anchors the rearmost selector fork pin, using locking fluid on its threads. Fit the stopper arm assembly, noting that its shouldered bolt also secures the retainer plate for the remaining selector fork pin. Hook the end of the stopper arm spring on the gearbox mainshaft bearing retainer, which has a suitable hole for this purpose.

33.1 Apply RTV sealant to jointing faces

33.2a Offer up the upper crankcase half

33.2b Lower crankcase bolt tightening sequence (X7)

33.2c Upper crankcase bolts (X7)

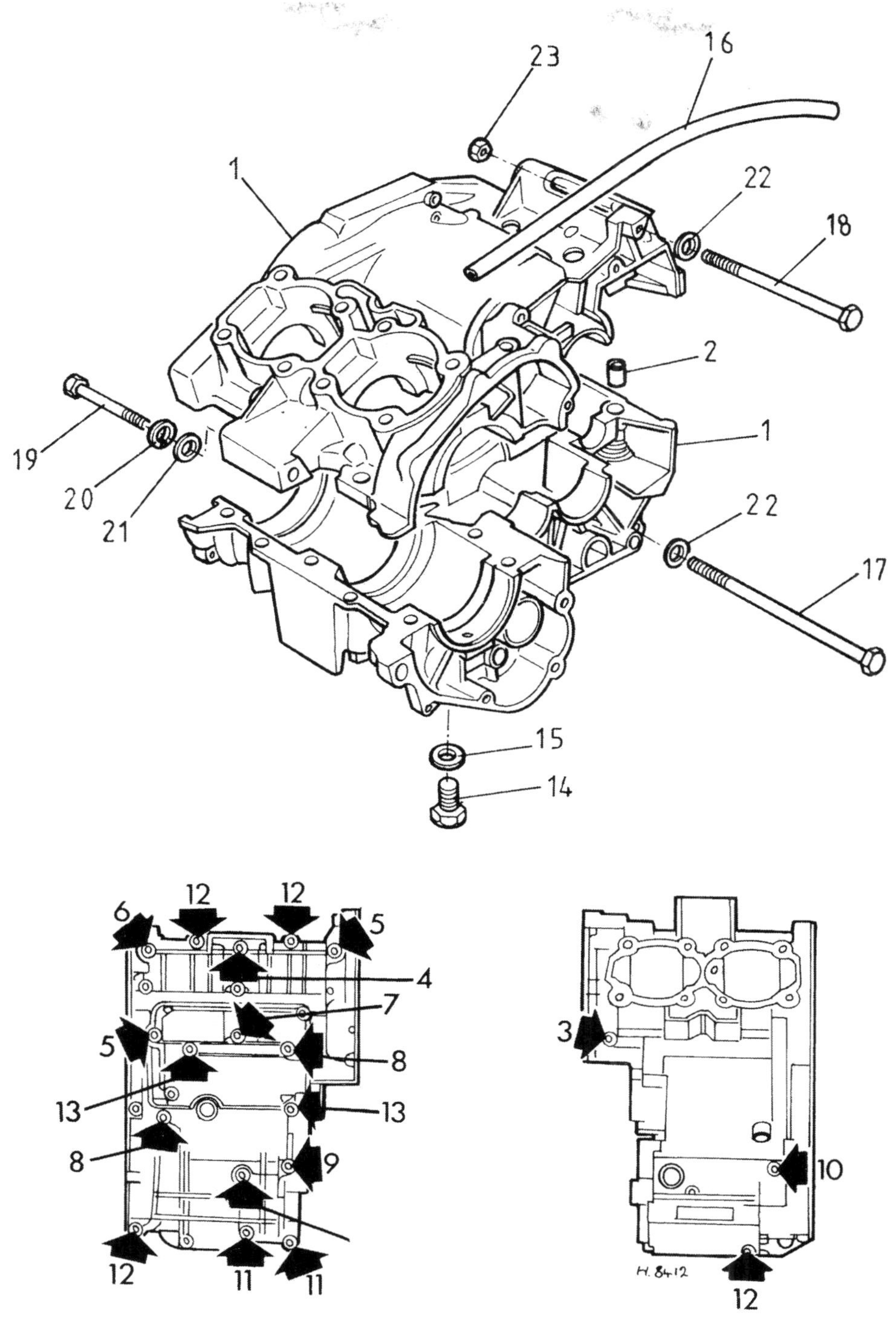

Fig. 1.21 Crankcases – SB200 and GT200 X5

1 *Complete crankcase assembly*
2 *Dowel pin – 2 off*
3 *Bolt (EN model)*
4 *Bolt*
5 *Bolt – 2 off*
6 *Bolt*
7 *Bolt*
8 *Bolt – 2 off*
9 *Bolt*
10 *Bolt*
11 *Bolt – 3 off*
12 *Bolt – 4 off*
13 *Bolt – 2 off*
14 *Drain plug*
15 *Drain plug gasket*
16 *Breather pipe*
17 *Engine mounting bolt*
18 *Engine mounting bolt*
19 *Bolt – 2 off*
20 *Spring washer – 2 off*
21 *Washer – 2 off*
22 *Washer – 2 off*
23 *Nut – 2 off*

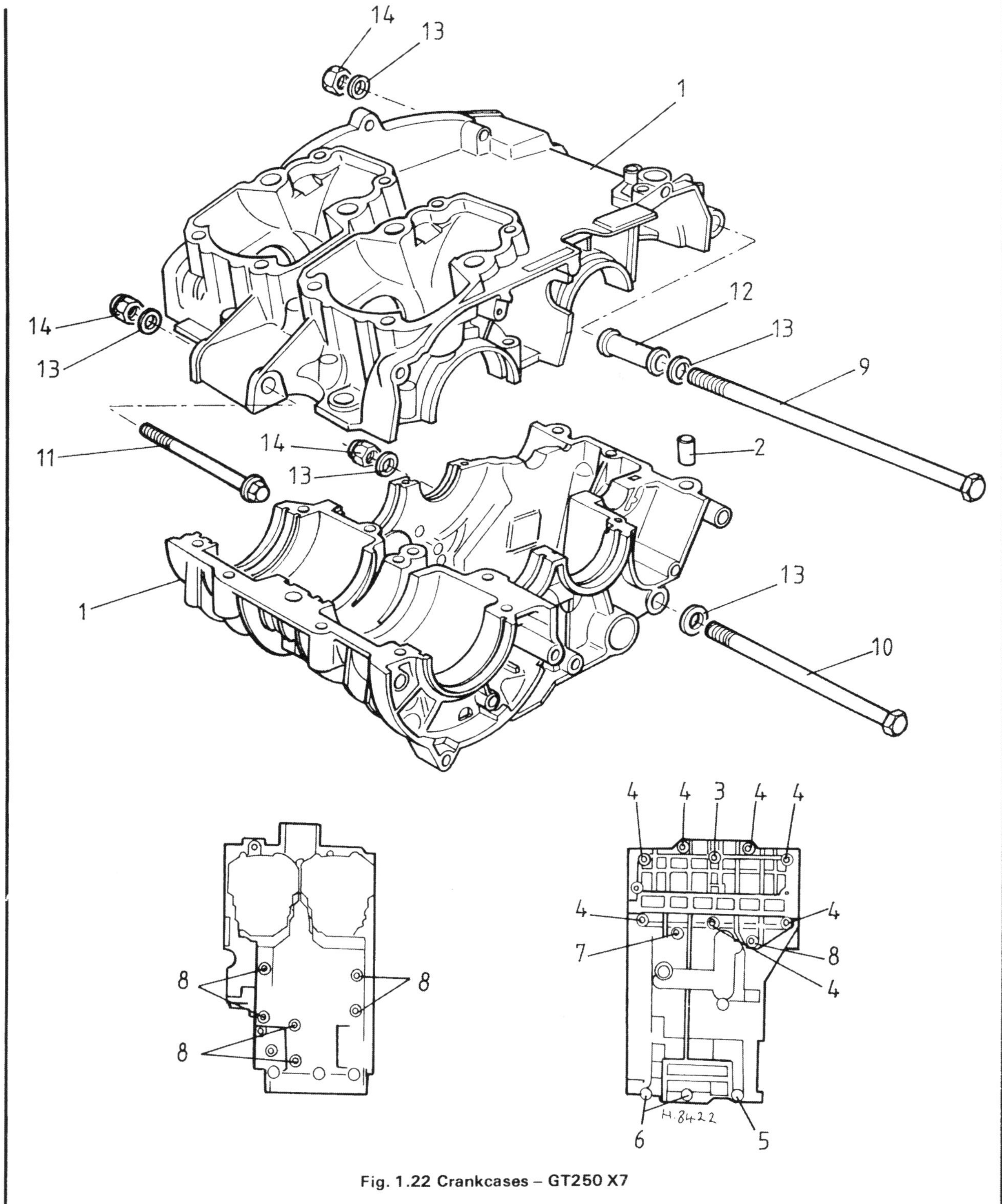

Fig. 1.22 Crankcases – GT250 X7

1 Complete crankcase assembly
2 Locating dowel – 2 off
3 Bolt
4 Bolt – 7 off
5 Bolt
6 Bolt – 2 off
7 Bolt
8 Bolt – 7 off
9 Engine mounting bolt
10 Engine mounting bolt – 2 off
11 Bolt
12 Engine mounting spacer
13 Washer – 7 off
14 Nut – 4 off

35 Engine reassembly: installing the gear selector mechanism – X7

1 With the engine unit inverted, refit the combined gearbox drain plug and neutral detent plunger assembly. It is important to ensure that the correct sealing washer is used, as any deviation in thickness will affect the operation of the plunger. Refit the gearbox mainshaft bearing retainer, using a thread locking fluid on the threads of its two securing screws.

2 Reassemble the selector ratchet assembly, noting that the slot in each ratchet pawl is slightly offset. It is important to fit the pawl with the slot innermost. Hold the pawls in against their springs, and insert the ratchet assembly into the recess in the end of the selector drum. The teeth on the outer face of the ratchet are grouped assymetrically, there being five teeth in one group and four in the other. Arrange the ratchet so that the group of five teeth are facing rearward, in line with the selector shaft bore.

3 Refit the selector ratchet retainer plate, and the pawl guide plate, each of which is secured by two screws. Fit the gearchange centring spring locating pin, which doubles as a selector fork shaft retainer. All the above threads should be treated with a thread locking fluid prior to fitting. Check that the selector shaft centring spring is fitted correctly, with the spring ends parallel rather than crossed, then slide the shaft into position. When fitted correctly, an imaginary line drawn between the centres of the selector shaft and the ratchet, should pass through the centre tooth of the ratchet and through the centre of the spring locating pin.

4 It is a good idea to check the gear selection at this stage. Temporarily refit the gearchange pedal, and attempt to select each gear in turn. It will be necessary to rotate the projecting layshaft end to ease engagement. The gears should select up and down the range.

36 Engine reassembly: installing the kickstart mechanism – X5 and SB200

1 The bulk of the kickstart mechanism used on the GT200 X5 and SB200 models is contained within the crankcase, and consequently will have been refitted prior to the crankcase halves being joined. It now remains to refit the return spring and spring guide. Check that the kickstart shaft is turned fully clockwise to its normal resting position.

2 Hook the end of the return spring in its hole in the lower casing. Grasp the straight tang, which projects inwards from the spring coils, using a pair of pointed-nose pliers. Turn the spring anti-clockwise by about 90° until the tang can be inserted through the hole in the shaft end. Refit the spring guide to hold the spring central to the shaft and to prevent the tang from becoming displaced.

35.1a Fit combined drain plug/detent bolt

35.1b Assemble bearing retainer, using locking fluid on screws

35.2a Slot in pawls must be offset *inwards*

35.2b Selector ratchet installed in end of selector drum

35.2c Group of *five* teeth should face rearward. Fit guide and retainer plates, and locating pin

35.3a Slide selector shaft into place, engaging centring spring

35.3b Mechanism should align as shown here

37 Engine reassembly: installing the kickstart mechanism – X7

1 Place the spring, followed by the ratchet block, over the end of the kickstart shaft, noting the position of the alignment mark on the boss of the ratchet. The ratchet teeth must face outward. Temporarily refit the kickstart lever, and turn the shaft until the index mark on the shaft faces upwards. The ratchet should be fitted over the splines so that the two index marks coincide. Release the kickstart lever and check that the lug on the ratchet engages in the stop on the casing.

2 The kickstart pinion can now be slid into position, followed by the small tachometer drive pinion. This is followed by a thrust washer, a wave washer and a second thrust washer. Slide the kickstart idler gear over the projecting end of the gearbox layshaft. The idler gear is followed by a plain washer and a retaining circlip.

37.1a Refit the kickstart stop, and lock with tabwasher

37.1b Shaft is marked with alignment dot ...

37.1c ... which should match corresponding dot on ratchet

37.1d Turn shaft anticlockwise, align dots, and fit ratchet

37.1e When shaft is released, ratchet should engage in stop

37.2a Fit kickstart pinion and plastic tachometer drive

37.2b Note arrangement of plain and wave washers

37.2c Idler pinion is secured by thrustwasher and circlip

38 Engine reassembly: refitting the clutch and crankshaft pinion – X5 and SB200

1 Fit the large plain washer over the crankshaft end, then insert the Woodruff key in its cutout. Slide the crankshaft pinion into position, followed by the tachometer drive pinion and the belleville washer (fit as shown in Fig. 1.23). Secure the assembly with the nut, lock the crankshaft using the method employed on removal and tighten the nut to 4.0–6.0 kgf m (29.0–43.5 lbf ft).
2 Check that the clutch centre springs are fitted correctly. The inner ends of the springs should be screwed firmly into the clutch centre so that they are flush with, but do not protrude beyond, the inner face of the clutch centre. Slide the spacer over the end of the gearbox mainshaft end, then fit the clutch outer drum, thrust washer and clutch centre.
3 Refit the tab washer and clutch centre nut, then tighten the nut down using the same method of holding the clutch centre as was used during dismantling. The nut should be tightened to 3.0–5.0 kgf m (21.5–36.0 lbf ft). Do not forget to bend up the tab washer to secure the clutch centre nut.
4 Install the pushrod components, noting that the shorter of the two pushrods should be fitted nearest the clutch. The rounded end of each should face towards the clutch. Position the final, headed, push piece, and place the thrust bearing in position, having greased the rollers first.
5 Place the clutch cover in position, ensuring that the index mark on the cover is aligned with the line on the inner face of the clutch centre. The clutch springs can be tensioned by pulling them outwards with a pair of pointed nose pliers whilst the anchor pin is slipped into place. Alternativelly, a simple wire hook can be made up for this purpose, using a piece of stiff wire such as that used in wire coathangers.

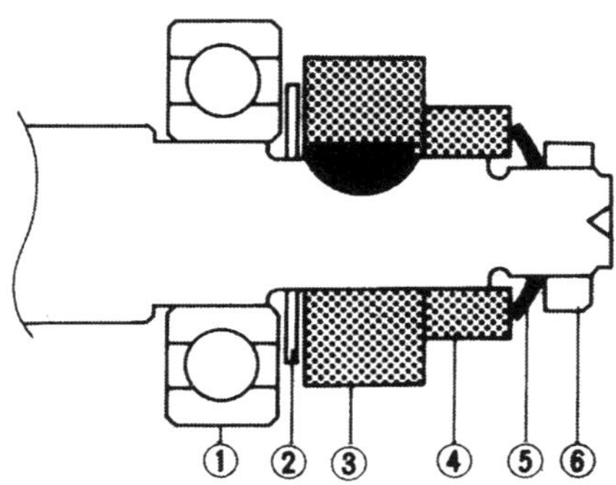

Fig. 1.23 Crankshaft pinion arrangement – GT200 X5

1 Bearing
2 Washer
3 Crankshaft pinion
4 Tachometer drive gear
5 Belville washer
6 Securing nut

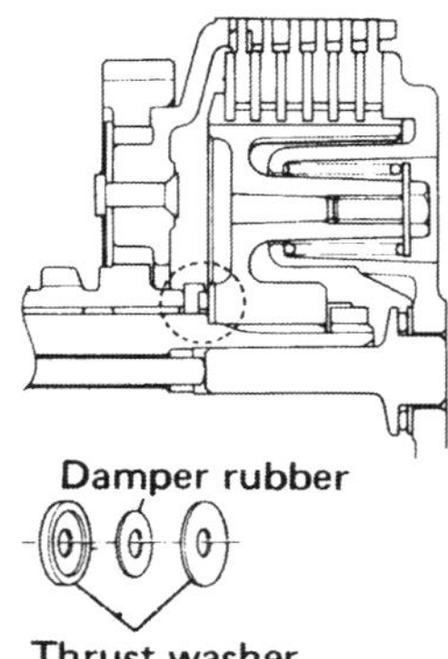

Fig. 1.24 Clutch thrustwasher arrangement – GT250 X7

39 Engine reassembly: refitting the clutch and crankshaft pinion – X7

1 Place the large plain washer over the end of the crankshaft, then fit the Woodruff key. Position the crankshaft pinion, tab washer and retaining nut, tightening the latter to 4.6–6.0 kgf m (29.0–43.5 lbf ft).
2 Place the large, plain, thrust washer in position over the end of the gearbox mainshaft, then fit the clutch outer drum. Fit the thick recessed thrust washer, the rubber washer, and the remaining thrust washer before offering up the clutch centre, tab washer and securing nut. Hold the clutch centre by whatever means was employed during removal, and tighten the clutch centre nut to 4.0–6.0 kgf m (29.0–43.5 lbf ft). Using pliers or similar, bend up one face of the tab washer to prevent the nut from slackening.
3 It will be noted that one of the five clutch friction plates differs from the rest. It has three recessed tangs, which locate small rubber blocks. These should be examined prior to installation, and new ones fitted if worn or damaged. When fitting new rubber blocks, ensure that the projection on each block faces in the same direction.
4 Fit the clutch plates in an alternating sequence, starting with a plain plate. The special friction plate is fitted next, with the projections facing inwards. Build up the remaining plates, finishing with a friction plate. Insert the two-piece clutch pushrod, noting that the shorter of the two rods should be fitted nearest the clutch, and that the rounded end of both should face towards the clutch. Assemble the clutch release bearing on the headed push piece, then fit the plain steel thrust washer. It is important that the bearing runs on the washer, and not on the light alloy clutch cover.
5 Place the clutch cover (pressure plate) in position, and fit the six springs, washers and retaining bolts. The bolts should be tightened progressively and evenly to preclude any possibility of warping. The final torque setting is 0.4–0.7 kgf m (3.0–5.0 lbf ft).

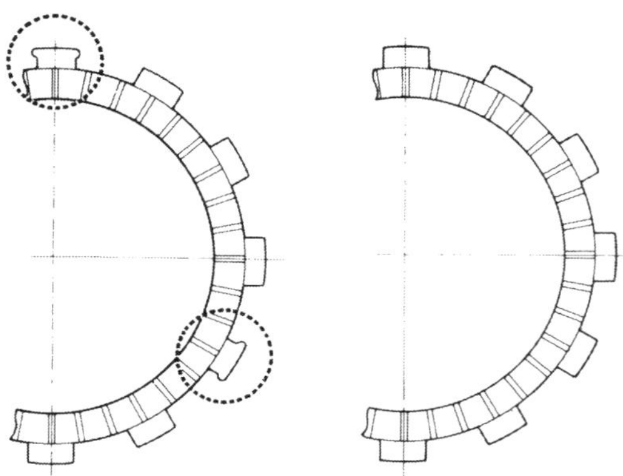

Fig. 1.25 Inner friction plate – GT250 X7

Circles denote special tangs for damping blocks

39.1a Fit large plain washer, Woodruff key and pinion

39.1b Lock crankshaft and secure retaining nut

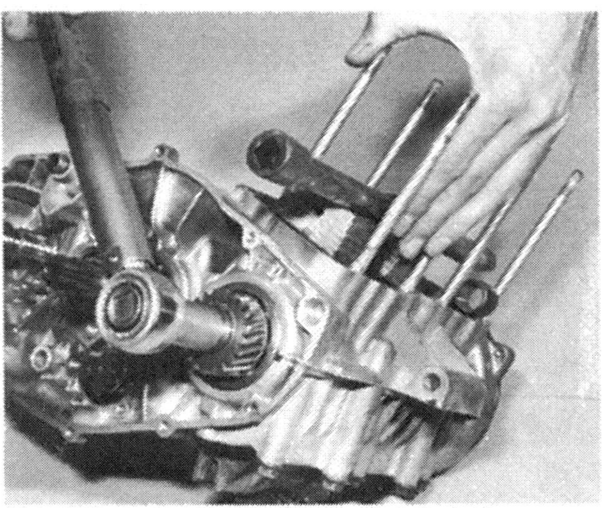
39.1c Safe method of preventing crankshaft rotation

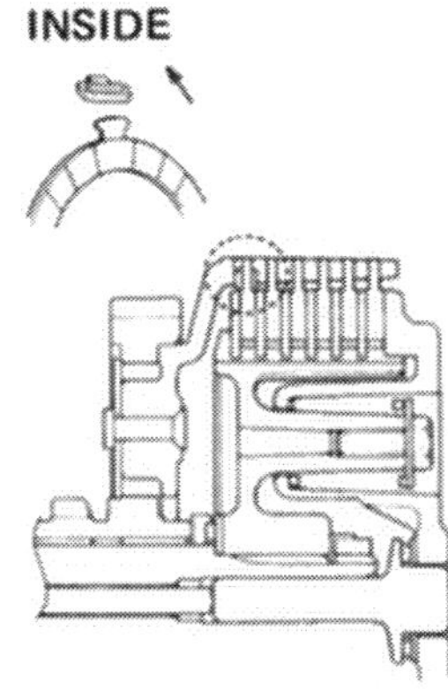

Fig. 1.26 Damping rubber location – GT250 X7

39.2a Fit thrustwasher to mainshaft end ...

39.2b ... and then offer up the clutch outer drum

39.2c Fit stepped washer, rubber washer and thrust washer

39.2d Clutch centre can now be fitted

39.2e Fit the clutch centre nut and tab washer

39.2f Lock the transmission, and tighten centre nut

39.3a Note damper rubbers on innermost friction plate (X7)

39.3b Damper rubbers should be fitted as shown

39.4 Fit clutch plates and assemble clutch release bearing

39.5a Assemble clutch springs, washers and bolts ...

39.5b ... tightening them down progressively

40 Engine reassembly: refitting the neutral switch and oil pump

1 Place the small compression spring and the neutral switch contact pin in the small hole in the left-hand end of the gear selector drum. Offer up the white plastic switch cover and secure it with two cross-head screws.

2 Check that the oil pump is clean and dry, and that the inlet and feed pipes are fitted securely over their relevant stubs on the pump body. Note that if the pipes show any signs of splitting or hardening, they should be renewed, as a failure of the lubrication would prove disastrous to the engine.

3 Fit a new paper gasket to the pump jointing face, and arrange the driving tang to coincide with the slot at the centre of the pump driving pinion. Thread the various pipes up through the aperture in the crankcase, then offer up the pump, securing it with its two cross-head screws. **IMPORTANT NOTE:** After the engine has been refitted and the oil supply connected, it is essential to bleed the pump before the engine is run. Serious damage may be caused if this operation is overlooked. See Section 45.6 of this Chapter for details.

40.1a Fit spring and switch plunger in selector drum ...

40.1b ... then secure switch cover with two screws

40.3 Refit the oil pump using a new gasket

4 Place the aforementioned thrust washer over the crankshaft end, fit the Woodruff key, and then offer up the assembled clutch and rotor assembly. Fit the lock washer and rotor securing nut, tightening it to 5.5–6.5 kgf m (40.0–47 lbf ft). The crankshaft can be prevented from turning by passing a stout bar through one of the small end eyes and supporting the ends of the bar on wooden blocks placed against the crankcase mouth.
5 If the alternator stator was removed from the outer casing, it should be refitted using a thread locking fluid on the retaining screws. Route the wiring through the recess in the casing, ensuring that the rubber grommet is correctly positioned. Note that the ignition timing is fixed on assembly, and subsequent adjustment is impossible on the 200 cc models.

SB200

6 The reassembly sequence is similar to that described above, with the obvious exception of the starter motor, which is fitted to the X5 only.

41 Engine reassembly: refitting the alternator and starter motor and drive – X5 and SB200

1 With the engine unit inverted, offer up the starter motor, positioning the shaft through the aperture in the casing. Fit the single retaining screw. Check that the starter motor lead is routed correctly and that the grommet is correctly placed where the lead passes through the casing. Refit the pressed steel cover, securing it with its four retaining bolts.
2 Working from the left-hand side of the unit, assemble the starter idler pinion on its support shaft, noting that a thrust washer is fitted at either end. Fit the assembled shaft and pinion into the casing, checking that the pinion meshes correctly with the starter motor shaft.
3 If the roller clutch has been removed from the back of the alternator rotor, refit it using thread locking fluid on the three Allen bolts which secure it. Fit the three springs, plungers and rollers into the clutch body and position the combined clutch boss and clutch gear. Before fitting this assembly over the end of the crankshaft, check that the double needle roller bearing is in position and note that a thrust washer should be fitted first.

42 Engine reassembly: refitting the alternator – X7

1 Offer up the alternator stator, noting that the scribed line on the stator should be in line with the centre of the upper securing screw. This should ensure that the timing is set satisfactorily, but this should be checked by using a stroboscopic timing lamp when the engine is started. See Chapter 3 for details. It should be noted that stroboscopic timing checks require the engine to be run at 6000 rpm, and as this is far from desirable with a newly rebuilt engine, the static setting should be regarded as correct until the engine has been run in.
2 Fit the Woodruff key in the crankshaft keyway, then place the alternator rotor in position. Fit the plain washer, spring washer and securing nut. The nut should be tightened to 4.5–7.0 kgf m (32.5–50.5 lbf ft). Suzuki recommend the use of a locking fluid on the crankshaft thread to obviate any tendancy for the nut to loosen in use.

42.1 Mark on stator should align with centre of screw

42.2a Fit Woodruff key ...

42.2b ... and offer up rotor

42.2c Lock crankshaft and tighten securing nut

43 Engine and gearbox reassembly: refitting the pistons and cylinder barrel(s)

1 Before commencing assembly of the pistons and cylinder barrels, pad the mouth of each crankcase with clean rag, to prevent any misplaced component from dropping in. Replace the needle roller bearings in the connecting rod eyes.

2 Fit each piston in turn, complete with rings. If the gudgeon pins are a tight fit in the piston bosses, warm the pistons first by placing a rag soaked in hot water on the crown. Note that the arrow mark on the piston crowns **must** face towards the exhaust port.

3 Use only new circlips and make sure each is positively located after the gudgeon pins have been pressed home. Never be tempted to re-use the original circlips; if they work loose as a result of having lost their tension serious damage is inevitable.

4 Make sure both small ends are well lubricated by using a hand oil can filled with engine oil. Pump some into the crankshaft and big-end assemblies too, before the cylinder barrels are fitted. This will necessitate removing the rag that pads each crankcase, which should be replaced until the pistons and piston rings have engaged with each cylinder bore.

5 Before refitting the cylinder barrel, oil the bore surfaces. If the cylinders have been rebored, check to ensure the edges of the ports have been rounded off correctly. Details are given in Section 23.8 of this Chapter. Fit new base gaskets (no cement).

6 Before inserting each piston into its respective cylinder barrel, check that the piston rings are aligned correctly with regard to their end pegs. Failure to observe this precaution will lead to ring breakages, apart from extreme difficulty in fitting.

7 The pistons should be replaced in their original position, as defined by the marks scratched on the inside of the skirt during the dismantling operation. Make sure that the arrow stamped on the crown of each piston faces the forward direction ie the front of the machine.

8 There is a generous taper on the base of each cylinder barrel (cylinder block, 200cc models) which should act as a good lead-in and obviate the need for a piston ring compressor. Feed each ring into the cylinder bore separately, using as little force as possible. When the pistons and rings are well up the cylinder bores, remove the rag padding from the crankcase mouths and lower the cylinder barrels until they seat firmly on their base gaskets.

9 Holding down the cylinder barrels by hand, rotate the engine a few times in both directions to make quite sure both pistons are running free in the bores. If everything appears in order, the cylinder head can be fitted.

43.1 Lubricate small end bearings prior to installation

43.2 Arrow on pistons must face forward

43.3 Use *new* circlips to avoid risk of failure

43.8 Cylinder holding bolts are fitted via holes in fins

44 Engine reassembly: refitting the cylinder head

1 Place a new cylinder head gasket over the projecting studs, noting that the layered gasket used on the X7 must be fitted as shown in the accompanying photograph (44.1), the plain side facing upwards. Place the cylinder head over the holding down studs, fit the plain washers, and fit the eight cylinder head nuts finger-tight.

2 Tighten each nut gradually, in the sequence shown in the accompanying line drawing, pulling the cylinder head down evenly to a final torque figure of 2.0–2.5 kgf m (14.5–18.0 lbf ft). Refit the various rubber damping blocks between the cylinder head fins.

3 Refit the sparking plugs, if only as a temporary expedient, to prevent foreign matter from dropping into the engine whilst it is being replaced. Do not over-tighten; it is sufficient to obtain a good seating between the plugs, their sealing washers and the cylinder heads, without excessive force.

44.1 On X7 model, plain face of head gasket faces upwards

44.2a Refit cylinder head, tightening bolts in sequence

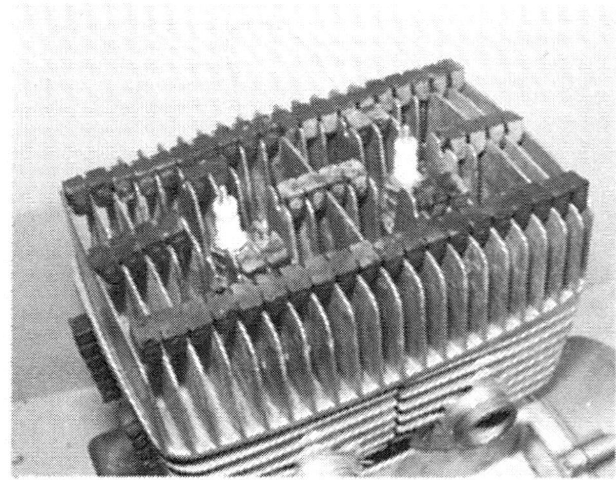

44.2b Do not omit rubber damping blocks between fins

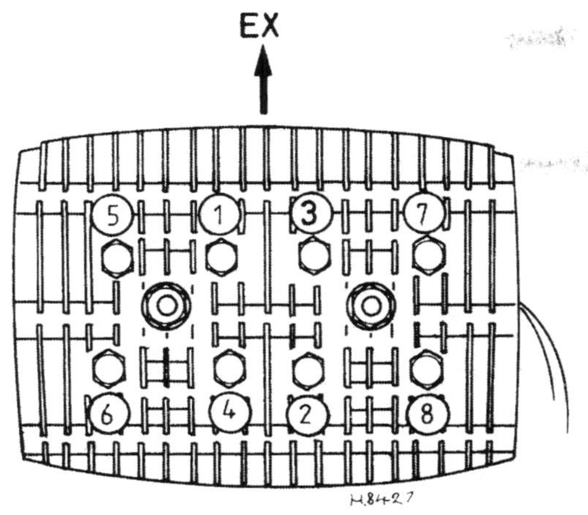

Fig. 1.27 Cylinder head tightening sequence – GT250 X7

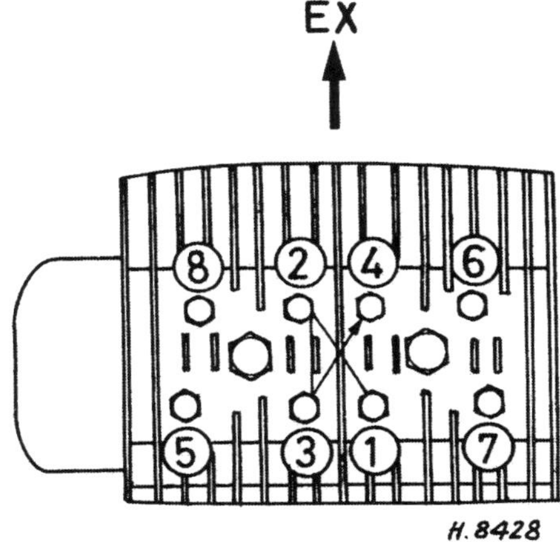

Fig. 1.28 Cylinder head tightening sequence – GT200 X5 and SB200

45 Refitting the engine/gearbox unit in the frame

1 It is helpful to enlist the aid of a convenient bystander at this stage, as this will aid the early stages of installation considerably. If essential, the unit can be lifted by one person without incurring permanent damage to the machine or owner. The unit is best fitted from the right-hand side, taking care not to topple the machine in the process. The unit can be rested conveniently on the lower frame tubes.

2 On X7 machines, a spacer is fitted between the frame and the engine unit, concentric to the rear engine mounting bolt. This must be fitted as the engine mounting bolts are positioned, ensuring that the drive chain (if of the endless type) is looped around it. It may prove necessary to tap the spacer into position. Assemble the engine mounting bolts and engine front plates loosely, preferably with new self-locking nuts. The smaller, 8 mm, nuts on the X5 and SB200 should be tightened to 1·8 – 2·3 kgf m (13·0 – 16·5 lbf ft), and larger 10 mm nuts, and all nuts on the X7, to 2·5 – 4·0 kgf m (18·0 – 29·0 lbf ft).

3 Reconnect the rear brake light switch operating spring, and refit the tachometer drive cable (X7 and X5 only), securing it with its gland nut. Refit the sparking plug caps, noting which is left and right, as marked during removal. Refit the silencers, and offer up the exhaust pipes. The latter are secured to the cylinder barrel by a retainer plate and two bolts and spring washers. A new gasket should be used at each joint. Tighten the clamp which secures each exhaust pipe to its silencer.

4 Reconnect the alternator output leads at the multi-pin connector. The connector is assymetrical, making incorrect connections impossible. Check that the alternator harness is routed correctly through the guide clips on the crankcase. Reconnect the feed pipe and operating cable to the oil pump. Note that the oil pump must be bled and adjusted before the machine is used, but after the carburettors have been fitted.

5 Offer up the carburettors as a pair, engaging each end of the choke in its rubber connecting hose. This procedure, whilst not impossible, calls for considerable persistence, and it may prove easier to release the screws holding the air filter housing so that more manoeuvring room is available. Do not omit to position the retaining clips over the adaptor hoses **before** the carburettors are fitted. Tighten the retaining clips, then refit the throttle valve assemblies, ensuring that the valves and needles are correctly located before screwing down the carburettor tops.

6 Referring to Chapter 2, check that the carburettors are correctly adjusted, and that the throttle cable free play is correct, then synchronise the carburettor slide and the pump alignment marks. The oil pump should also be bled, to remove any air from the pipe between the tank and the pump body. Slacken and remove the bleed screw, which can be identified by its fibre sealing washer, and allow the oil to flow through the pump until all air bubbles have been flushed out. When the oil pump adjustment and bleeding have been completed, refit the pressed steel cover.

7 Engage the gearbox sprocket on the chain and slide it into position on the layshaft end. On all X7 models, also early X5 and SB200 models, fit the retaining plate and the tab washer, then apply a few drops of thread locking compound to their threads and refit the two bolts, tightening them to a torque setting of 0.6 – 1.0 kgf m (4.5 – 7.5 lbf ft). Bend up the two locking tabs of the tab washer against the head of each bolt to secure it. On later X5 and SB200 models, fit the lockwasher then apply thread locking compound to the layshaft threads and refit the large retaining nut, tightening it to a torque setting of 4.0 – 6.0 kgf m (29 – 43 lbf ft). Bend up an unused portion of the lockwasher against one of the flats of the nut to secure it. Note that these large nuts have been known to slacken in spite of the lock washer, therefore the use of thread locking compound is highly advisable. Working as described in Routine Maintenance, check that the drive chain is correctly adjusted and fully lubricated.

8 If the ignition timing is to be checked (X7 only), carry out the instructions in paragraphs 9 and 10 below so that the engine can be started before the left-hand engine casing is refitted. On refitting the cover (all models) check that the clutch pushrod is correctly inserted and that it engages fully with the release mechanism. Securely tighten the cover retaining screws and check that the clutch cable is routed correctly, with no sharp twists. Working as described in Routine Maintenance, adjust the clutch so that the cable has 4 mm (0.16 in) of free play on X5 and SB200 models and 2 – 3 mm (0.08 – 0.12 in) on X7 models. Check that the clutch is operating correctly before taking the machine out on the road.

9 Refit the battery, and reconnect the battery leads, taking great care to observe the correct polarity; negative (–) earth. The reu lead **must** be connected to the positive (+) terminal. Refit the fuel tank, and reconnect the petrol feed pipe to the tap. Remove the gearbox oil level plug, which is to be found on the right-hand engine casing, to the rear of the Suzuki emblem on X7 models, and to the front of this on X5 and SB200 models. Remove the filler plug and add SAE 20W/40 engine oil until it reaches the threads of the level hole. The X5 and SB200 will require about 700 cc (1·2/1·4 US/Imp pint) whilst the X7 will accept about 800 cc (1·69/1·41 US/Imp pint) Refit the level screw and filler plug.

10 Make a final check around the machine to ensure that no connections have been missed. Before riding the machine, check the normal adjustments, chain and brakes etc, and ensure that the electrical system functions normally. Refit the kickstart and gearchange pedals, checking that they are in a convenient position when sitting astride the machine.

45.2 Engine mounting spacer is a tight fit

45.7a Do not forget to refit clutch pushrod

45.7b Refit sprocket and lock securing bolts

Fig. 1.29 Gearbox sprocket mounting GT200 X5 and SB200 – later models only

46 Starting and running the rebuilt engine

1 When the initial start-up is made, run the engine slowly for the first few minutes, especially if the engine has been rebored or a new crankshaft fitted. Check that all the controls function correctly and that there are no oil leaks, before taking the machine on the road. The exhausts will emit a high proportion of white smoke during the first few miles, as the excess oil used whilst the engine was reassembled is burnt away. The volume of smoke should gradually diminish until only the customary light blue haze is observed during normal running. It is wise to carry a spare pair of sparking plugs during the first run, since the existing plugs may oil up due to the temporary excess of oil.

2 Remember that a good seal between the pistons and the cylinder barrels is essential for the correct functioning of the engine. A rebored two-stroke engine will require more carefully running-in, over a longer period, than its four-stroke counterpart. There is a far greater risk of engine seizure during the first hundred miles if the engine is permitted to work hard.

3 Do not tamper with the exhaust system or run the engine without baffles fitted to the silencer. Unwarranted changes in the exhaust system will have a very marked effect on engine performance invariably for the worse. The same advice applies to dispensing with the air cleaner or the air cleaner element.

4 Do not on any account add oil to the petrol under the mistaken belief that a little extra oil will improve the engine lubrication. Apart from creating excess smoke, the addition of oil will make the mixture much weaker, with the consequent risk of overheating and engine seizure. The oil pump alone should provide full engine lubrication.

47 Fault diagnosis: engine

Symptom	Cause	Remedy
Engine will not start	Defective spark plugs	Remove plugs and lay them on cylinder head. Check whether spark occurs when engine is kicked over.
	Ignition timing set incorrectly	Check and reset (see Chapter 3).
	Discharged battery	Check whether lights work. If battery is flat, remove and charge.
	Air leak at crankcase or worn crankshaft oil seals	Flood carburettors and check whether petrol is reaching the plugs.
Engine runs unevenly	Ignition and/or fuel system fault	Check as though engine will not start.
	Blowing cylinder head gasket	Oil leak should provide evidence. Renew gasket.
	Incorrect ignition timing	Check and if necessary adjust.
	Carburettors out of balance	Refer to Chapter 2 and adjust.
	Choked silencers	Remove baffles and clean.
Lack of power	Incorrect ignition timing	See above.
	Fault in fuel system	Check system and vent in filler cap.
	Choked silencers	See above.
White smoke from exhaust	Too much oil	Check oil pump setting.
	Tank contains two-stroke petroil and not straight petrol	Drain and refill with straight petrol.
Engine overheats	Pre-ignition and/or weak mixture	Check carburettor settings, also grade of plugs fitted.
	Lubrication failure	Stop engine and check oil pump setting. Is oil tank dry?

48 Fault diagnosis: clutch

Symptom	Cause	Remedy
Engine speed increases but machine does not respond	Clutch slip	Check whether clutch adjustment still has free play. Check thickness of linings and renew if near wear limit.
Difficulty in engaging gears, gear changes jerky and machine creeps forward, even when clutch is withdrawn fully	Clutch drag	Check clutch adjustment to eliminate excess play. Check whether clutch centre and outer drum have indented slots.
Operating action stiff	Clutch assembly loose on mainshaft	Check tightness of retaining nut.
	Bent pushrod	Renew.
	Dry pushrod	Lubricate.
	Damaged, trapped or frayed control cable	Check cable and renew if necessary. Make sure cable is lubricated and has no sharp bends.

49 Fault diagnosis: gearbox

Symptom	Cause	Remedy
Difficulty in engaging gears	Selector forks or rods bent	Renew.
	Broken springs in gear selector mechanism	Check and renew.
	Clutch drag	See preceding Section.
Machines jumps out of gear	Worn dogs on ends of gear pinions	Strip gearbox and renew worn parts.
	Sticking camplate plunger	Remove plunger cap and free plunger assembly.
Kickstarter does not return	Broken return spring	Remove right-hand crankcase cover and renew spring.
Kickstarter slips or jams	Worn ratchet assembly	Remove right-hand crankcase cover, dismantle kickstarter assembly and renew worn parts.
Gear change lever does not return	Broken return spring	Remove right-hand crankcase cover and renew spring.

Chapter 2 Fuel system and lubrication

Contents

Specifications

Carburettors	GT250X7	GT200EN	GT200EX	SB200
Make	Mikuni	Mikuni	Mikuni	Mikuni
Type	VM26SS	VM22SS	VM22SS	VM18SS
Bore size	26 mm	22 mm	22 mm	18 mm
ID number	11300 – EN 11301 – EX	10300	10301	10210
Main jet	87.5	82.5	82.5	70
Pilot jet	25	20	20	17.5
Needle jet	O–1	O–4	O–8	N–9
Jet needle	5CN21	4D25	4CG1	4F19
Needle clip position – grooves from top	3	4	3	3
Throttle valve cutaway	2.0	2.0	2.0	2.0
Pilot air screw – turns out	$1\frac{1}{4}$	$1\frac{1}{4}$	$1\frac{1}{4}$	$1\frac{1}{4}$
Float height	25.0 ± 1 mm	19.9 ± 1 mm	19.9 ± 1 mm	19.9 ± 1 mm

Engine lubrication	All models
Type	Pump fed total-loss system (Suzuki CCI)
Engine oil tank capacity:	
GT250 X7	1.5 lit (2.6 pint)
GT200 X5 and SB200	1.3 lit (2.3 pint)

1 General description

1 The fuel system comprises a petrol tank, from which petrol is fed by gravity to the float chambers of the twin Mikuni carburettors, via a three position petrol tap. The tap has 'Off', 'On' and 'Reserve' positions, the latter providing a warning that the petrol level is low in time for the owner to find a garage.

2 The carburettors are of conventional concentric design, the float chambers being integral with the lower part of the carburettor bodies. Cold starting is assisted by a separate starting circuit which supplies the correct fuel-rich mixture when the 'choke' control is operated. The two are interconnected by a common linkage, and are operated by a single lever on the left-hand instrument.

3 Air entering the carburettors passes through a moulded plastic air cleaner casing, which contains an oil-impregnated foam air filter. This effectively removes any airborne dust, which would otherwise enter the engine and cause premature wear. The air cleaner also helps silence induction noise, a common problem inherent with two-stroke engines.

4 Engine lubrication is catered for by the Suzuki CCI system. Oil from a separate tank is fed by an oil pump to small injection nozzles in the inlet tract. The pump is linked to the throttle twistgrip, and this controls the volume of oil fed to the engine. Unlike the previous system employed by Suzuki, the oil is not pressure-fed to the various moving parts. Instead, the oil mist is drawn in with the fuel/air mixture, and is deposited inside the engine to provide lubrication of the crankshaft components and the pistons and cylinder walls.

2 Petrol tank: removal and replacement

1 It is unlikely that the petrol tank will need to be removed except on very infrequent occasions, because it does not restrict access to the engine unless a top overhaul is to be carried out whilst the engine is in the frame.
2 The petrol tank is secured at the rear by a single bolt, washer and rubber buffer that threads into a strut welded across the two top frame tubes. It is necessary first to remove the dual seat before access is available.
3 When the bolt and washer are withdrawn, the petrol tank can be lifted clear from the frame. The nose of the tank is a push fit over two small rubber buffers, attached to a peg that projects from each side of the frame, immediately to the rear of the steering head. A small rubber 'mat' cushions the rear of the tank and prevents contact with the two top frame tubes.
4 The petrol tank has a locking filler cap to prevent pilferage of the tank contents when the machine is left unattended. The lock is actuated by the ignition key.
5 Unlike most two-strokes, there is no dependence on the use of a petrol/oil mix for lubrication of the engine. The engine lubrication system is quite separate and in consequence only petrol alone is carried in the petrol tank.

3 Petrol tap: removal, dismantling and reassembly

1 The petrol tap is secured to the underside of the tank by two hexagon-headed screws. There is seldom need to disturb the main body of the petrol tap. In the event of a leak at the operating lever, the complete lever assembly can be dismantled (provided the petrol tank is drained first) with the main body undisturbed.
2 The tap lever assembly can be withdrawn for inspection, after releasing the single crosshead screw which locates it. If leakage has been evident, the most likely culprit will be the O-ring which seals the tap valve against the body. If this fails to effect a cure, it will be necessary to renew the complete tap assembly.
3 The tap is provided with a sediment bowl, in which any fine debris from the tank which has managed to get through the filter gauze, will be trapped, along with any water. The bowl should be periodically removed for cleaning. When refitting the sediment bowl, ensure that the O-ring is in good condition.
4 Before reassembling the petrol tap, check that all parts are clean, especially the tube which forms the filter and main and reserve intakes. A new gasket should be fitted to the filter bowl assembly in order to effect a satisfactory seal.
5 Do not overtighten any of the petrol tap components during reassembly. The castings are in a zinc-based alloy, which will fracture easily if over-stressed. Most leakages occur as the result of defective seals.

4 Petrol feed pipes: examination

The petrol feed pipes, connecting the carburettors to the fuel tap, are made of thin walled plastic, and are retained by small wire clips. Check that the pipes have not split or become brittle, due to age and the effects of heat and fuel. Check also the various drain and breather pipes.

5 Carburettors: general description

1 The Mikuni twin carburettors fitted to the Suzuki X7, X5 and SB 200 twins are of the Amal type, using a conventional throttle slide and needle arrangement that works in conjunction with a main jet to control the amount of petrol/air mixture administered to the engine.
2 Air is drawn into the carburettor bell mouths via a large capacity air cleaner that has an oil impregnated foam element. The air cleaner acts also as an effective carburettor intake silencer and eliminates induction 'roar'. The engine must not be run without the air cleaner attached because the carburettors are jetted to compensate for the restriction in air flow. Removal of the air cleaner will result in a greatly weakened mixture, which will cause overheating and subsequent engine damage.

6 Carburettors: removal

1 If the removal of either carburettor is necessary, it will be easier to remove and replace them as a pair, as there is a cold start mechanism linkage, and a short pipe connecting the two which balances the inlet depression in each inlet tract. It is advisable, in any case, to work on both instruments simultaneously, as they are matched as a pair and should correspond closely in operation. Never dismantle or disturb the carburettors needlessly, as balancing the refitted units must be precise, and requires fine adjustment, which can be time-consuming.
2 Detach the feed pipes from the float chamber of each carburettor by displacing the small wire clip and sliding it up the pipe. This is best done by squeezing together the small ears of the clip with a pair of pointed-nose pliers. Do not pull the pipes with undue force as this will strain and possibly break the plastic. Ease them gently upwards with the aid of a small electrical screwdriver.
3 Remove the short connecting hoses between each of the carburettor air intakes, and the studs on the air filter trunking. These are each retained by two screwed clips. Detach breather and drain tubes in the same manner as the petrol feed pipes were removed.
4 Unscrew the carburettor tops, and withdraw the slides complete with needles and return springs. Unless these need specific attention, tie them clear by taping them to the frame top tube.
5 Loosen the clip which clamps the carburettors to the rubber inlet stubs, and draw the two carburettors, with their balance pipe, clear. Note the locating pips and corresponding notches in the carburettors and rubber stubs. Place the carburettors on a bench for further dismantling.

3.1 Tap is secured to underside of tank by two bolts

3.2 Tap lever can be removed after releasing screw

3.3 Sediment bowl unscrews for cleaning

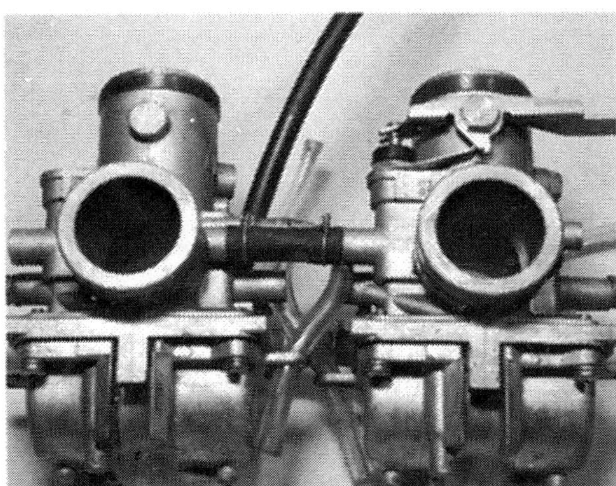
6.5 Carburettors are removed as a pair

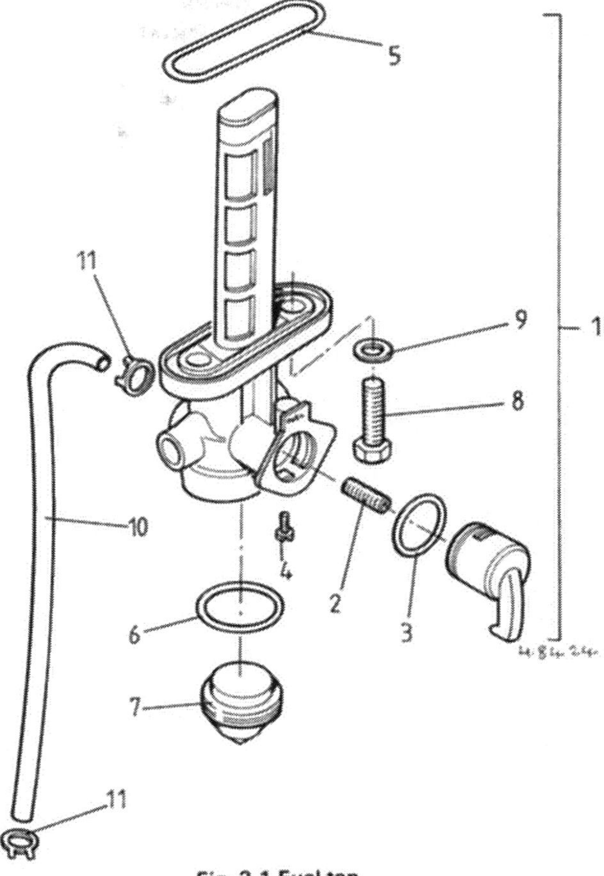

Fig. 2.1 Fuel tap

1 Complete tap assembly
2 Spring
3 O-ring
4 Screw
5 O-ring
6 Filter bowl gasket
7 Filter bowl
8 Bolt – 2 off
9 Gasket – 2 off
10 Fuel pipe
11 Clip – 2 off

7 Carburettors: dismantling and reassembly

1 Invert each carburettor and remove the float chamber by withdrawing the four retaining screws. The float chamber bowls will lift away, exposing the float assembly, hinge and float needle. There is a gasket between the float chamber bowl and the carburettor body which need not be disturbed unless it is leaking.

2 With a pair of thin nose pliers, withdraw the pin that acts as the hinge for the twin floats. This will free the floats and the float needle. Check that none of the floats have punctured and that the float needle and seating are both clean and in good condition. If the needle has a ridge, it should be renewed in conjunction with its seating.

3 The two floats are made of plastic, connected by a brass bridge and pivot piece. If either float is leaking, it will produce the wrong petrol level in the float chamber, leading to flooding and an over-rich mixture. The floats cannot be repaired successfully, and renewal will be required.

4 The main jet is located in the centre of the circular mixing chamber housing. It is threaded into the base of the needle jet and can be unscrewed from the bottom of the carburettor. The needle jet lifts out from the top of the carburettor, after the main jet has been unscrewed.

5 The float needle seating is also found in the underside of the carburettor, towards the bell mouth intake. It is secured by a small retainer plate and is sealed by an O-ring. If the float needle and the seating are worn, they should both be replaced, never separately. Wear usually takes the form of a ridge or groove, which may cause the needle to seat imperfectly.

6 The carburettor slides, return springs and needle assemblies together with the mixing chamber tops, are attached to the throttle cable. The throttle cable divides into two at a junction box located within the two top frame tubes. There is also a third cable, which is used to link the oil pump with the throttle.

7 After an extended period of service the throttle slides will wear and may produce a clicking sound within each carburettor body. Wear will be evident from inspection, usually at the base of the slide and in the locating groove. Worn slides should be replaced as soon as possible bacause they will give rise to air leaks which will upset the carburation.

8 The needles are suspended from the slides, where they are retained by a circlip. The needle is normally suspended from the centre grooves, but other grooves are provided as a means of adjustment so that the mixture strength can be either increased or decreased by raising or lowering the needle. Care is necessary when replacing the carburettor tops because the needles are easily bent if they do not locate with needle jets.

9 The manually operated chokes are unlikely to require attention during the normal service life of the machine. When the plungers are depressed, fuel is drawn through a special starter jet in each carburettor by the partial vacuum that is created in the crankcases. Air from the float chamber passes through holes in the starter emulsion tube to aerate the fuel. The fuel then mixes with air drawn in via the starter air inlet to the plunger chamber. The resultant mixture, enrichened for a cold start, is drawn into the engine through the starter outlet, behind the throttle valve.

10 Before the carburettors are reassembled, using the reversed dismantling procedure, each should be cleaned out thoroughly, preferably by the use of compressed air. Avoid using a rag because there is always risk of fine particles of lint obstructing the internal air passages or the jet orifices.

11 Never use a piece of wire or any sharp metal object to clear a blocked jet. It is only too easy to enlarge the jet under these circumstances and increase the rate of petrol consumption. Always use compressed air to clear a blockage; a tyre pump makes an admirable substitute when a compressed air line is not available.

12 Do not use excessive force when reassembling the carburettors because it is quite easy to shear the small jets or some of the smaller screws. Before attaching the air cleaner hoses, check that both throttle slides rise when the throttle is opened.

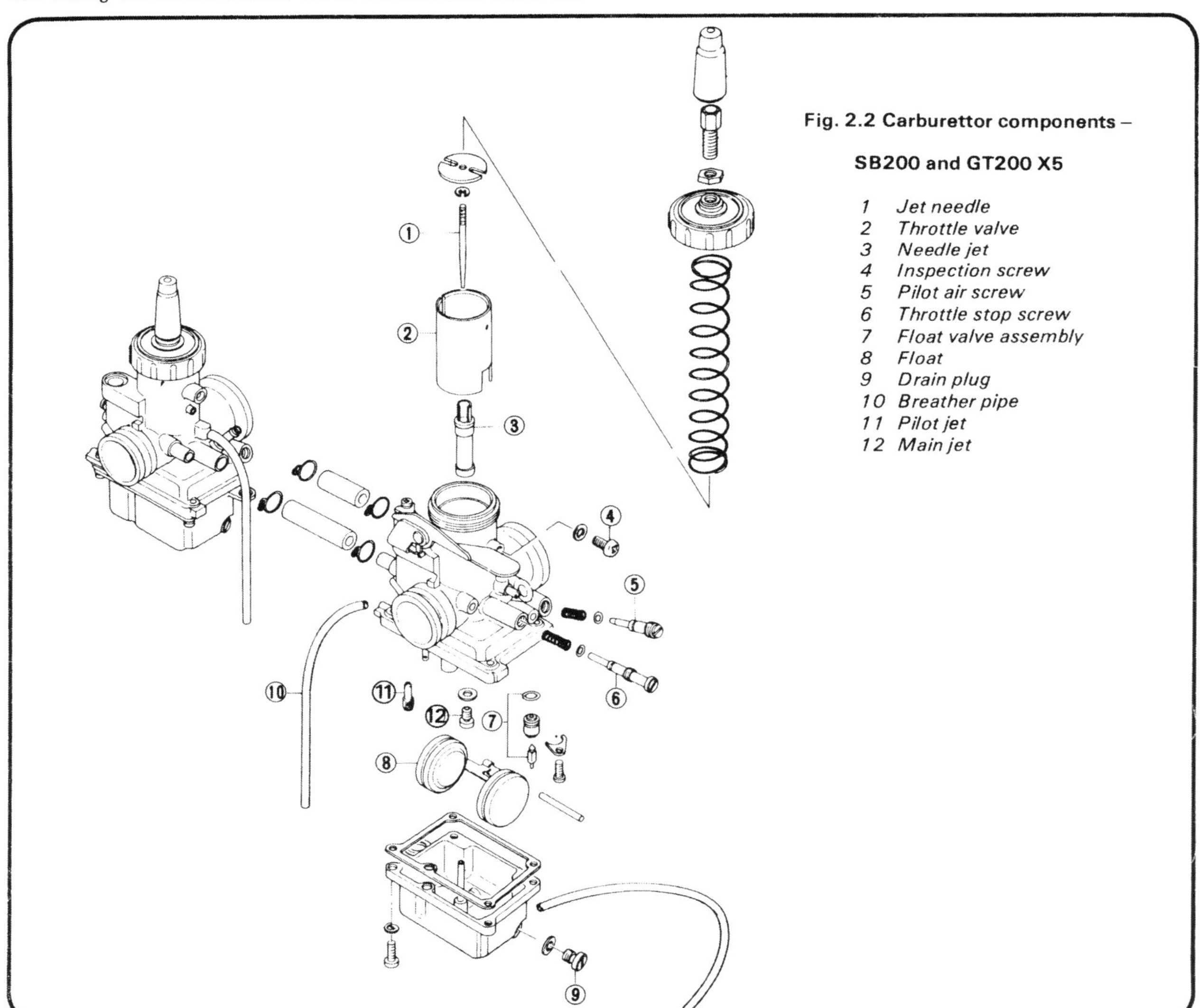

Fig. 2.2 Carburettor components – SB200 and GT200 X5

1 Jet needle
2 Throttle valve
3 Needle jet
4 Inspection screw
5 Pilot air screw
6 Throttle stop screw
7 Float valve assembly
8 Float
9 Drain plug
10 Breather pipe
11 Pilot jet
12 Main jet

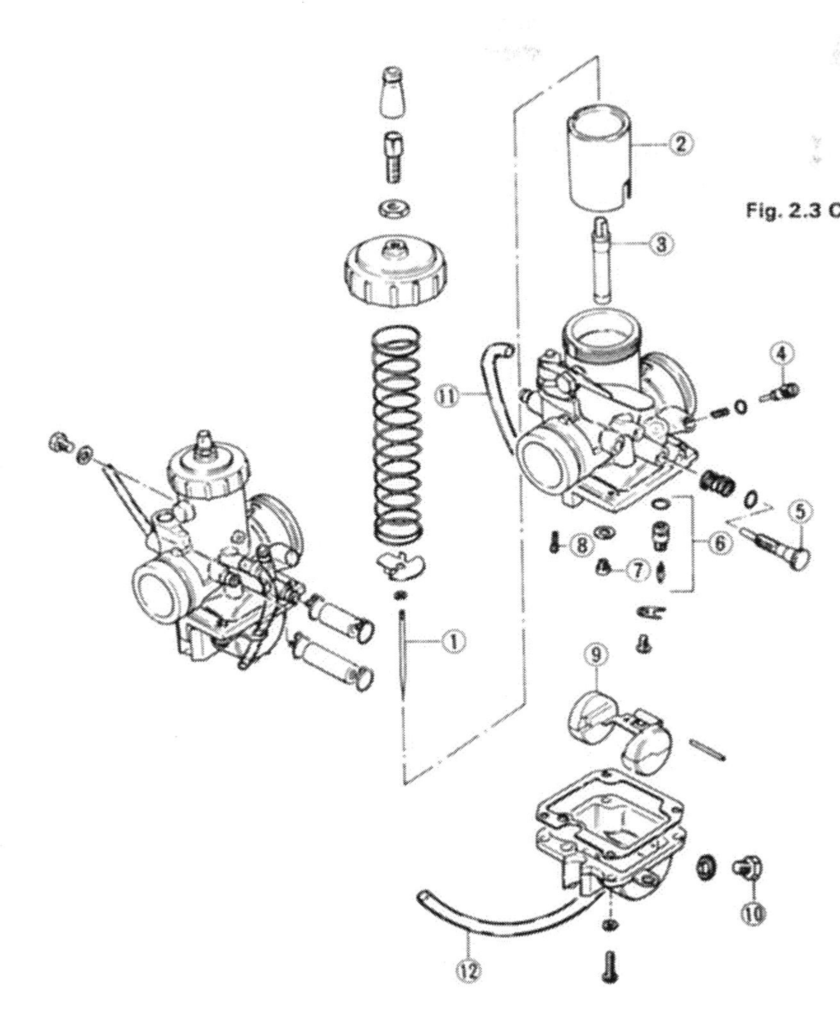

Fig. 2.3 Carburettor components – GT250 X7

1 *Jet needle*
2 *Throttle valve*
3 *Needle jet*
4 *Pilot air screw*
5 *Throttle stop screw*
6 *Needle valve assembly*
7 *Main jet*
8 *Pilot jet*
9 *Float*
10 *Drain plug*
11 *Breather hose*
12 *Overflow hose*

7.1 Float bowl is secured by four screws

7.2a Withdraw pivot pin ...

7.2b ... and lift float and needle away

7.4 Main jet screws into extended boss

7.5a Float needle seat is retained by plate and screw ...

7.5b ... and is fitted with an O-ring seal

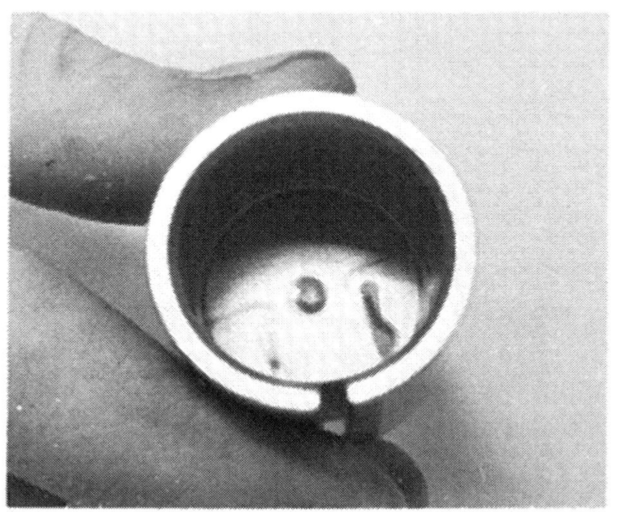
7 8a Remove retaining plate to allow ...

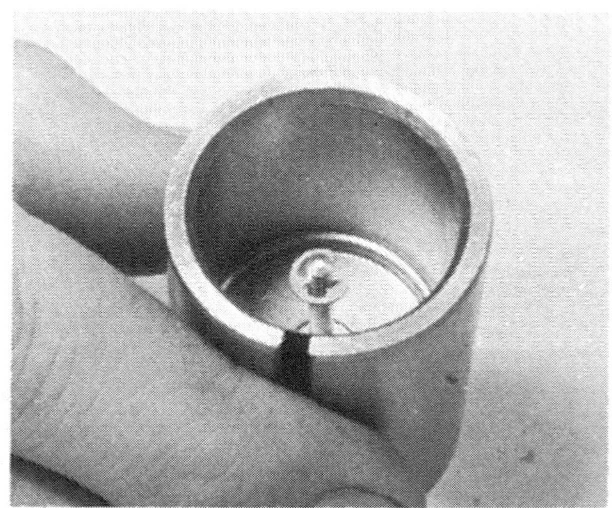
7.8b ... needle to be displaced from valve

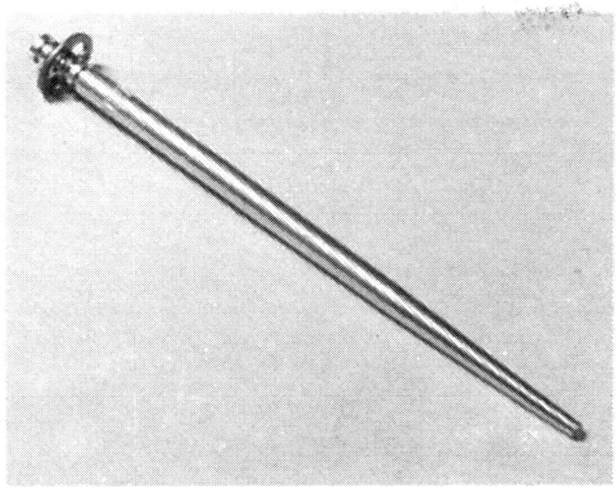

7.8c Needle grooves facilitate adjustment

7.9a Choke lever is retained by a shouldered bolt

7.9b Choke plunger screws into carburettor body

8 Carburettors: adjustment

1 When adjustments are required it is best to regard each carburettor as a separate entity. Remove the sparking plug from the cylinder that is not receiving attention so that only the one cylinder will fire during the adjustments. Then change over and follow a similar routine. These adjustments should be made with the engine warm.

2 Commence operations by checking the float level, which will involve detaching the carburettor concerned, inverting it and removing the float chamber bowl. If the float level is correct the distance between the uppermost portion of the floats and the flange of the mixing chamber will be as follows:

Model	Carburettor type	Float height
GT 250 X7	*Mikuni VM26SS*	*25 ± 1mm*
GT 200 X5	*Mikuni VM 22SS*	*19.9 ± 1 mm*
SB 200	*Mikuni VM18SS*	*19.9 ± 1 mm*

Adjustments are made by bending the tang on the float arm in the direction required (See accompanying diagram).

3 Replace the carburettor and turn the pilot air screw until it is closed fully. Then turn it back approximately 1 – 1¼ turns. Adjust the amount of play in the throttle cable to within 0.5 – 1 mm and then start the engine.

4 Readjust the pilot air screw until the engine runs smoothly at a stabilised engine speed. Note that a two-stroke engine will never fire evenly at very low engine speeds, so the setting will take the form of the best compromise. It will be necessary to adjust the throttle stop screw in conjunction with the pilot air screw, to achieve a stabilised engine speed. Fig. 2.2 and 2.3 show the location of these adjusting screws.

5 Stop the engine, replace the sparking plug and lead and then repeat the procedure with the other cylinder and carburettor after removing the sparking plug from the cylinder that has just been adjusted.

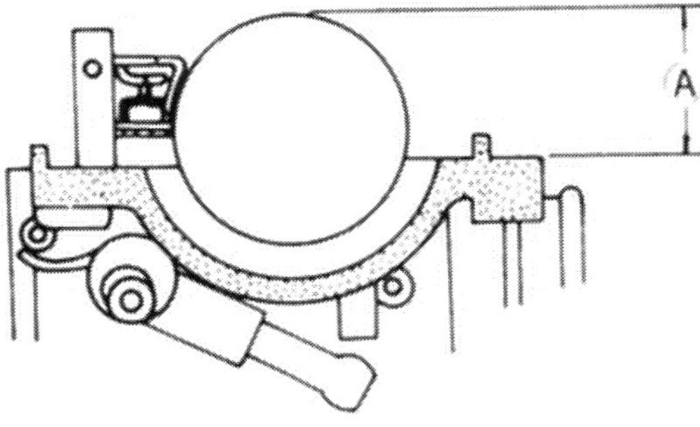

Fig. 2.4 Float height adjustment

Measure distance A, which should correspond with specified float height setting

9 Synchronizing the carburettors

1 Power output will be unbalanced unless both carburettors work in perfect harmony with each other. Many cases of poor performance and low power output can be traced to carburettors that are out of phase with each other.
2 It is imperative to check that both carburettor slides enter the bore of the carburettor at the same time. Remove the air cleaner hoses and open the throttle wide (engine dead) so that both slides are raised to their maximum height. Slowly close the throttle and check that both slides enter the carburettor bores at the same time. If they do not, make adjustments with either cable adjuster until they are in phase. Check that both slides close fully when the throttle is shut. Reconnect the air filter hoses and run the engine.
3 If carburettor adjustments are required, they should be made **before** the carburettors are synchronized. Always make the synchronizing adjustments with the engine warm, to obviate the risk of a false setting.

10 Carburettor settings

1 Some of the carburettor settings, such as the sizes of the needle jets, main jets and needle positions are predetermined by the manufacturer. Under normal circumstances it is unlikely that these settings will require modification, even though there is provision made. If a change appears necessary, it can often be traced to a developing engine fault.
2 As a rough guide, the pilot air screw controls the engine speed up to $\frac{1}{8}$ throttle. The throttle valve cutaway controls the engine speed from $\frac{1}{8}$th to $\frac{1}{4}$ throttle and the position of the needle from $\frac{1}{4}$ to $\frac{3}{4}$ throttle. The main jet is responsible for the engine speed at the final $\frac{3}{4}$ to full throttle. It should be added that none of these demarkation lines is clearly defined; there is a certain amount of overlap between the carburettor components involved.
3 Always err on the side of a rich mixture because a weak mixture has a particularly adverse effect on the running of any two-stroke engine. A weak mixture will cause rapid overheating which may eventually promote engine seizure. Reference to Chapter 3 will show how the condition of the sparking plugs can be used as a reliable guide to carburettor mixture strength.

11 Exhaust system: cleaning

1 The exhaust system is often the most neglected part of any two-stroke despite the fact that it has quite a pronounced effect on performance. It is essential that the exhaust system is inspected and cleaned out at regular intervals because the exhaust gases from a two-stroke engine have a particularly oily nature which will encourage the build-up of sludge. This will cause back pressures and restrict the engine's ability to 'breathe'.
2 Cleaning is made easy by fitting the silencers with detachable baffles, held in position by a set screw that passes through each silencer end. If the screw is withdrawn, the baffles can be drawn out of position for cleaning.
3 A wash with a petrol/paraffin mix will remove most of the oil and carbon deposits, but if the build-up is severe it is permissible to heat the baffles with a blow lamp and burn off the carbon and old oil.
4 At less frequent intervals, such as when the engine requires decarbonising, it is advisable also to clean out the exhaust pipes. This will prevent the gradual build-up of an internal coating of carbon and oil, over an extended period.
5 Do not run the machine with the baffles detached or with a quite different type of silencers fitted. The standard production silencers have been designed to give the best possible performance whilst subduing the exhaust note. Although a modified exhaust system may give the illusion of greater speed as a result of the changed exhaust note, the chances are that performance will have suffered accordingly.
6 When replacing the exhaust system, use new sealing rings at the exhaust port joints and check that the baffle retaining screws are tightened fully in the silencer ends.

12 Air cleaner: dismantling, servicing and reassembling

1 Before access can be gained to the air cleaner box and hoses it is necessary first to remove the front section of the left-hand side panel. On the 200cc models, the cover is secured by a single screw at the top, whilst on X7 models, it is retained by two small bolts at the front edge.
2 With the cover removed, the filter element on X5 and SB200 models can be released by disengaging the frame from its base. On the X7, release the external wing nut to allow the element to be withdrawn. The two types of element differ in design, but are similar in principle. Pull the two halves of the support frame apart (X5 and SB200) and remove the foam element. On X7 models, peel the foam element band off the edge of the filter frame.
3 Clean the foam element in a non-flammable solvent, squeezing it gently to dislodge any accumulated debris. Do not attempt to wring the foam element out as it is easily torn. The solvent should be removed by squeezing the foam carefully in some clean rag. Allow any residual solvent to evaporate.
4 Check the cleaned element carefully, looking for holes or splits, which if found, will necessitate the element's renewal. Immerse the cleaned element in engine oil, then squeeze it gently to remove any excess. The object is to leave it moist, but not dripping, with oil.
5 Do not run the machine with the air cleaner detached, even for a short period. The carburettors are jetted to compensate for the addition of the air cleaner and the carburation will be upset badly if it is removed. The effect will be overheating as a direct result of the permanently-weakened mixture strength and the risk of subsequent engine damage.
6 The air cleaner is reassembled by reversing the dismantling procedure. If the rubber hoses are split or perished, renew them in order to preclude the possibility of air leaks.

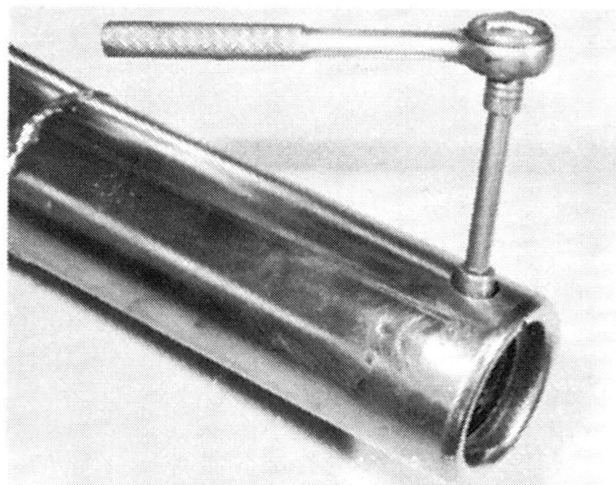

11.2a Remove bolt from silencer end ...

11.2b ... to permit baffle removal and cleaning

12.1 End cover is secured by two bolts (X7)

12.2a Filter element is retained by external wing nut (X7)

12.2b Withdraw the filter assembly ...

12.2c ... and remove foam element for cleaning

13 The lubrication system

1 Unlike early two-strokes, the Suzuki X7, X5 and SB200 models have an independent lubrication system for the engine and do not require the mixture of a measured quantity of oil to the petrol content of the fuel tank in order to utilise the so-called 'petroil' method. Oil of the correct viscosity is contained in a separate oil tank mounted on the left-hand side of the machine and is fed to a mechanical oil pump on the left-hand side of the engine which is driven from the gearbox by reduction gear. The pump delivers oil at a predetermined rate, via two flexible plastic tubes, to oilways in the inlet passage of each cylinder barrel. In consequence, the oil is carried into the engine by the incoming charge of petrol vapour, when the inlet port opens.

2 The oil pump is also interconnected by the twist grip throttle, so that when the throttle is opened, the oil pump setting is increased a similar amount. This technique ensures that the lubrication requirements of the engine are always directly related to the degree of throttle opening. This facility is arranged by means of a control cable attached to a lever on the end of the pump; the cable is joined to the throttle cable junction box at the point where the cable splits into two for the operation of each carburettor.

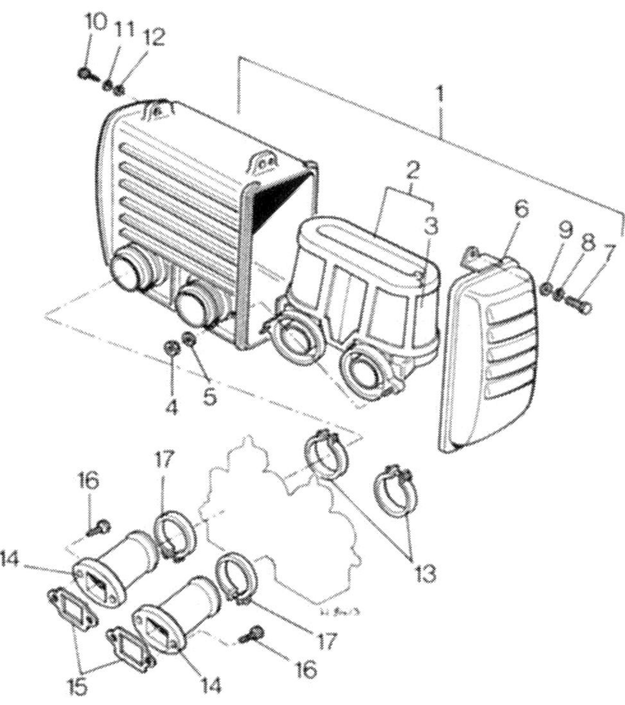

Fig. 2.5 Air filter

1 *Air filter assembly*
2 *Element frame*
3 *Element*
4 *Nut – 2 off*
5 *Washer – 2 off*
6 *Air filter case end cap*
7 *Screw – 2 off*
8 *Spring washer – 2 off*
9 *Washer – 2 off*
10 *Screw – 2 off*
11 *Spring washer – 2 off*
12 *Washer – 2 off*
13 *Clamp – 2 off*
14 *Outlet pipe – 2 off*
15 *Outlet pipe gasket – 2 off*
16 *Screw – 4 off*
17 *Clamp – 2 off*

14 Removing and replacing the oil pump

1 It is rarely necessary to remove the oil pump unless specific attention to it is required. The pump is located in a recess in the left-hand crankcase, and may be reached after releasing the left-hand outer casing. On X5 and SB200 machines, only the rearmost section of the casing need be released.
2 The pump is housed at the rear of the crankcase, and is protected by a pressed steel cover. It may be found helpful to remove the final drive chain to allow better access, although this is not essential. The pressed steel pump cover is secured by two screws.
3 Release the two screws which secure the pump body to the crankcase. Prise off the delivery pipes at the inlet manifolds. The feed pipe from the oil tank should be disconnected, noting that it will be necessary to plug the tank outlet or to drain its contents into a clean container. Pull the pump body clear, disengaging the oil pipes from the crankcase.
4 Further dismantling is not practicable, and it will be necessary to renew the pump if it is obviously damaged. Maintenance must be confined to keeping the pump clear of air, and correctly adjusted, as described in the following sections.
5 Refit the oil pump to the crankcase cover, using a new gasket at the oil pump/crankcase cover joint. Replace and tighten the two crosshead mounting screws. The remainder of the reassembly is accomplished by reversing the dismantling procedure, but do not replace the crankcase cover because the oil pump must be bled to ensure the oil lines are completely free from air bubbles. See the following Section.

13.1 Oil injection pipe in inlet port

6 Note that a modified type of pump is fitted as standard to all later SB200 and X7 models; this later type differs only in that the banjo unions are integral with the pump body. In all other respects the two types are identical and either type may be supplied as a replacement part should renewal be necessary.

15 Bleeding the oil pump

1 It is necessary to bleed the oil pump every time the main feed pipe from the oil tank is removed and replaced. This is because air will be trapped in the oil line, no matter what care is taken when the pipe is removed.
2 Check that the oil pipe is connected correctly, with the retaining clip in position. Then remove the cross-head screw in the outer face of the pump body with the fibre washer beneath the head. This is the oil bleed screw.
3 Check that the oil tank is topped up to the correct level, then place a container below the oil bleed hole to collect the oil that is expelled as the pump is bled. Allow the oil to trickle out of the bleed hole, checking for air bubbles. The bubbles should eventually disappear as the air is displaced by fresh oil. When clear of air, refit the bleed screw. DO NOT replace the front portion of the crankcase cover until the pump setting has been checked, as described in the next Section.

16 Checking the oil pump and throttle cable settings

1 As has been mentioned previously, the oil pump is controlled by the throttle twistgrip, and it is important that the pump lever is kept synchronized with the carburettor throttle valve. It should be noted that the oil pump setting must always be checked whenever the throttle cable play is altered, as each adjustment will affect the relationship with the other.
2 Check that the carburettors are correctly adjusted and synchronized as described earlier in this chapter, then set the throttle cable adjusters to give 1.0-1.5 mm (0.04-0.06 in) free play. Hold this setting, and secure each of the adjuster locknuts.
3 Remove the inspection plug from the side of the left-hand carburettor to enable the throttle valve to be viewed. Slowly open the throttle until a circular index mark can be seen. Align this so that it **just** touches the upper edge of the inspection hole. At this setting, the line on the oil pump lever should coincide exactly with its index mark. If this is not the case, alter the oil pump cable adjuster setting until alignment is achieved. Open and close the throttle a few times, then recheck the setting.

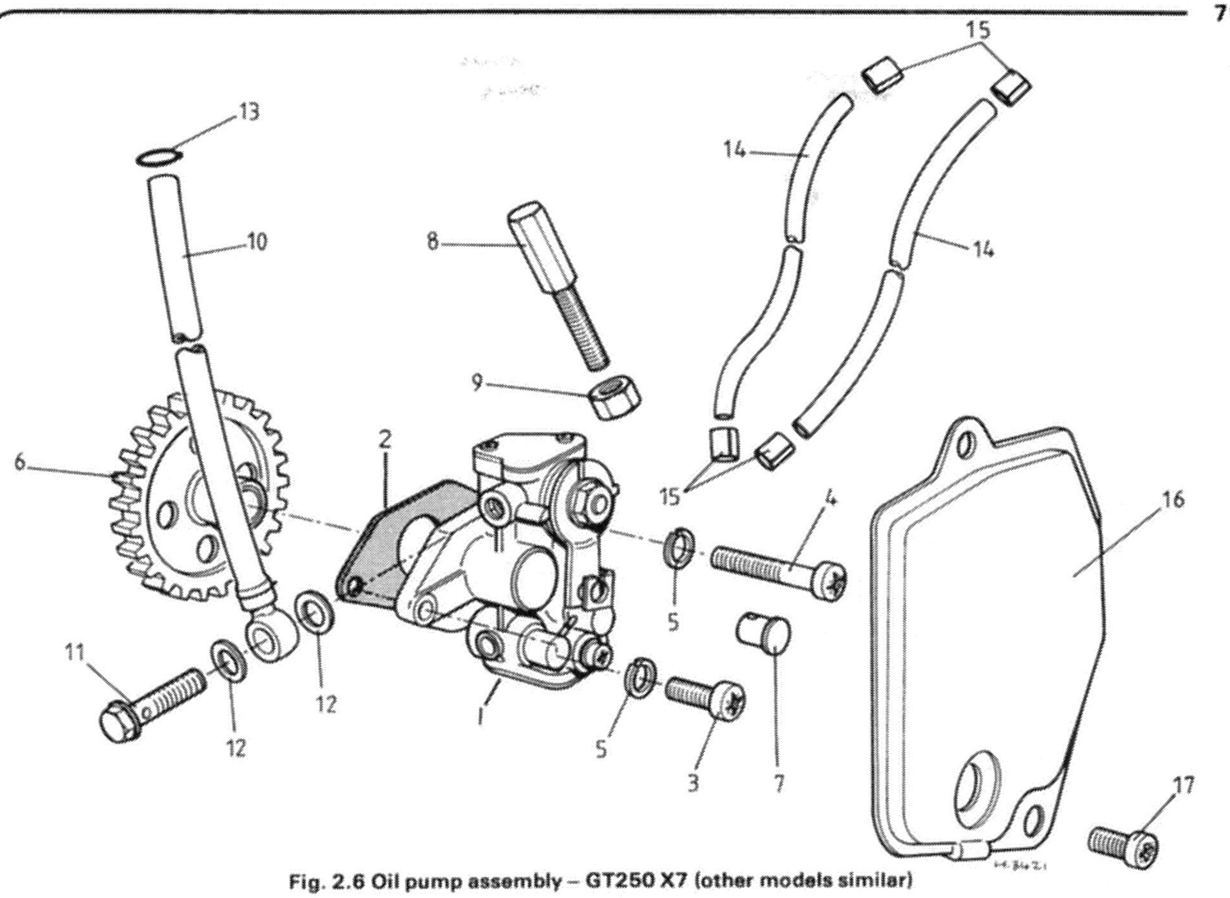

Fig. 2.6 Oil pump assembly – GT250 X7 (other models similar)

1 *Oil pump assembly*
2 *Oil pump gasket*
3 *Screw*
4 *Screw*
5 *Spring washer – 2 off*
6 *Oil pump drive pinion*
7 *Cable trunnion*
8 *Cable adjusting screw*
9 *Nut*
10 *Oil pump feed pipe*
11 *Union bolt*
12 *Gasket – 2 off*
13 *Clip*
14 *Output feed pipe – 2 off*
15 *Cable clamp – 4 off*
16 *Oil pump inspection cover*
17 *Screw – 2 off*

15.2 Pump is bled by removing this screw

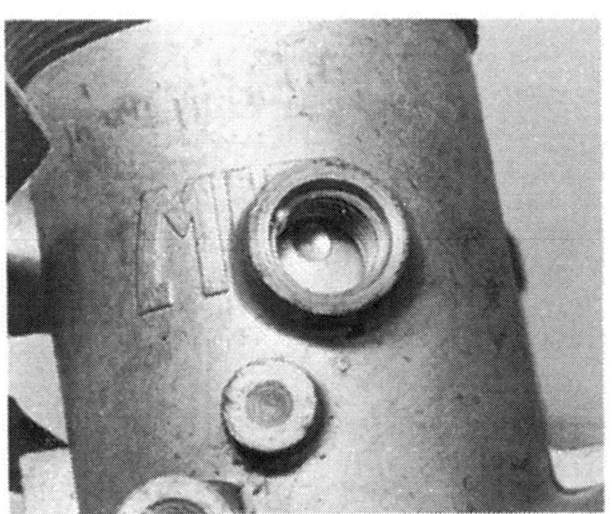

16.3a Alignment mark on throttle valve should be here ...

16.3b ... when pump alignment marks coincide

16.3c The oil pump alignment marks

17 Fault diagnosis: fuel system

Symptom	Cause	Remedy
Excessive fuel consumption	Air cleaner choked or restricted	Clean or renew element.
	Fuel leaking from carburettor	Check all unions and gaskets.
	Badly worn or distorted carburettors	Renew.
	Carburettor settings incorrect	Readjust. Check settings with specifications.
Idling speed too high	Throttle stop screw in too far	Adjust screws.
	Carburettor tops loose	Tighten.
Engine sluggish. Does not respond to throttle	Back pressure in silencers	Check baffles and clean if necessary.
Engine dies after running for a short while	Blocked vent hole in filler cap	Clean.
	Dirt or water in carburettors	Remove and clean.
General lack of performance	Weak mixture; float needle sticking in seat	Remove float chambers and check needle seatings.
	Air leak at carburettors or leaking crankcase seals	Check for air leaks or worn seals.

18 Fault diagnosis: lubrication system

Symptom	Cause	Remedy
White smoke from exhausts	Too much oil	Check oil pump setting and reduce if necessary.
Engine runs hot and gets sluggish when warm	Too little oil	Check oil pump setting and increase if necessary.
Engine runs unevenly, not particularly responsive to throttle openings	Intermittent oil supply	Bleed oil pump to displace air in feed pipe.
Engine dries up and seizes	Complete lubrication failure	Check for blockages in feed pipe, also whether oil pump drive has sheared.

Chapter 3 Ignition system

Contents

Specifications

	GT205 X7		GT 200 X5		SB200	
Ignition system						
Make			Suzuki PEI			
Type			Electronic			
Timing	20° BTDC @ 6000 rpm		24° BTDC @ 6000 rpm		24° BTDC @ 6000 rpm	
Ignition coil resistance:						
Primary	1 ohm		1 ohm		1 ohm	
Secondary	10 K ohms		20–30 K ohms		20–30 K ohms	
Sparking plug						
Make	NGK	ND	NGK	ND	NGK	ND
Type:						
Standard	B9ES	W27ES	B8ES	W24ES	B8ES	W20EP
Hot*	B8ES	W24ES	B7ES	W22ES	–	–
Cold*	B10E	W31ES	B9ES	W27ES	–	–
Gap	0.6–0.8 mm (0.024–0.032 in)					

Note: Changes of spark plug heat range from standard will not normally be required if the machine is functioning correctly. Holed pistons may result from indiscriminate substitution of plugs.

1 General description

1 The Suzuki X7, X5 and SB200 models are equipped with an electronic ignition system of the capacitor discharge, or CDI, type. Despite its rather unfortunate title, the Suzuki Pointless Electronic Ignition (PEI) system offers many significant advantages over conventional coil-and-contact breaker arrangements.

2 The most important advantage of electronic ignition, is that it removes all mechanical components from the system, the spark being triggered electronically by a pulser coil and magnet rather than by a contact breaker assembly. Because there are no contact breakers to wear, the owner is freed from the task of periodically adjusting or renewing them. Once the electronic system has been set up, it need not be attended to unless it has been disturbed in the course of dismantling or failure occurs in the electronic components in the system.

3 The ignition exciter, or source, coil assembly is housed within the flywheel generator, and provides the power supply for the external CDI unit, which is mounted beneath the right-hand side panel. The system is triggered by a magnet, incorporated in the outer face of the generator rotor, acting upon a pickup coil, known as a pulser, which is outrigged from the stator. The sparking plugs are supplied from a single ignition coil, firing in both cylinders simultaneously. This system is known as the 'spare spark' system, as only one cylinder can fire, leaving one spark wasted, or 'spare'.

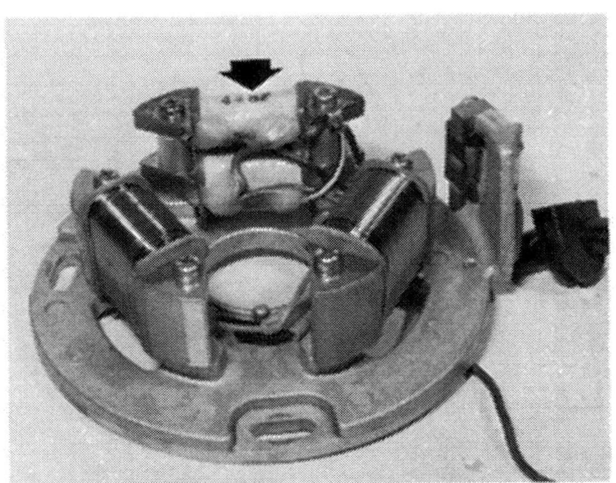

1.3a Ignition source, or exciter, coil, is located on stator plate

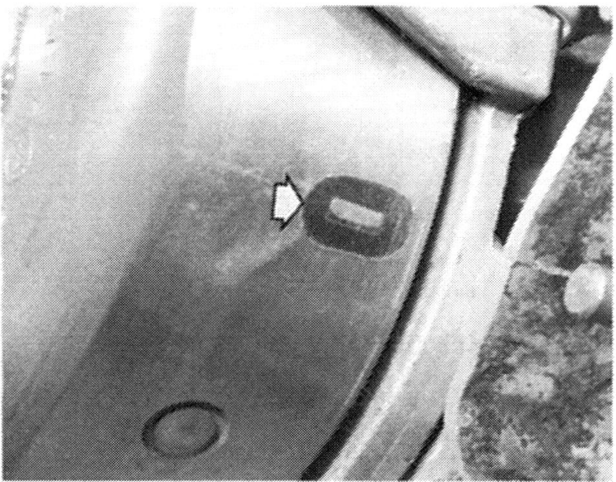
1.3b Magnet set in rotor periphery ...

1.3c ... triggers the outrigged pulser coil

2 Electronic ignition system: principles of operation

1 Energy for the ignition system is drawn from the **exciter coil**. This is mounted on the generator stator, and is somewhat different in appearance to the two charging coils. It is, in fact, a two-stage arrangement, having low speed and high speed windings. The low speed windings produce a high output voltage at low engine speeds, this voltage dropping off as the engine builds up speed. The high speed windings, on the other hand, produce little energy at low engine speeds, but the output voltage rises along with engine speed. The two outputs are combined, offsetting each other to give a fairly constant output voltage, this being the sum of the output of each set of windings.

2 The exciter coil assembly feeds the **CDI unit**, a sealed electronic assembly which forms the heart of the system. This unit contains, amongst other things, a capacitor and a thyristor, or silicon controlled rectifier (SCR). The capacitor is charged with the high voltage output from the exciter coil assembly. The thyristor, or SCR, is in effect an electronic switch. When signalled electrically by the pulser, it allows the capacitor to discharge through the primary windings of the ignition coil. This in turn induces a high tension pulse in the secondary windings, which is fed to the sparking plugs.

3 The **pulser**, or pickup, comprises a small coil mounted outside the alternator rotor on a projection from the stator. A permanent magnet embedded in the flywheel rotor is arranged to pass beneath the pulser coil. As the magnet passes the pulser coil, a weak current is generated and it is this that is used to trigger the thyristor in the CDI unit.

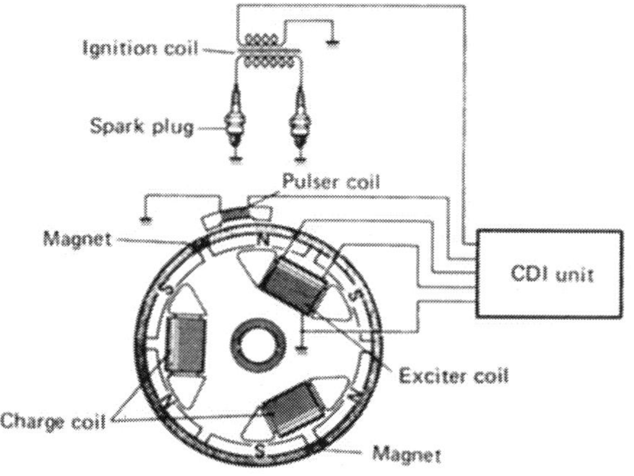

Fig. 3.1 The CDI ignition system

3 Electronic ignition system: testing and maintenance

1 As stated earlier in this Chapter, the electronic ignition needs no regular maintenance once it has been set up and timed accurately. Occasional attention should, however, be directed at the various connections in the system, and these must be kept clean and secure. A failure in the ignition system is comparatively rare, and usually results in a complete loss of ignition. Usually, this will be traced to the CDI unit, and little can be done at the roadside to effect a repair. In the event of the CDI unit failing, it must be renewed as repair is not practicable.

2 If the CDI unit is thought to be at fault, it is recommended that it be removed and taken to a Suzuki dealer for testing. The dealer will have the use of a special CDI tester, and will be able to test the unit accurately and quickly. Testing at home is less practical and cannot be guaranteed to be accurate. At best, it will enable the owner to establish which part of the system is at fault, although replacement of the defective part remains the only effective cure.

3 Care must be exercised when dealing with the CDI unit. Wrong connections could cause instant and irreparable damage to the unit, so this must be an important consideration when removing and replacing the unit. If the unit is to be disconnected and removed for testing, use a short length of insulated wire to short-circuit each pair of terminals to avoid any electric shock from residual energy in the capacitor.

4 For those owners possessing a multimeter and who are fully conversant with its use, a test sequence is given below. It is not recommended that the inexperienced attempt to test the system at home, as more damage could be sustained by the system, and an unpleasant shock experienced by the unwary operator. It should be appreciated that the CDI system is capable, in certain circumstances, of producing sufficient current output to be dangerous to the operator. It follows that an attitude to safety should be adopted similar to that applying when testing housefold mains electricity. Before performing any tests, the following points must be noted. Disconnect all the connectors in the ignition system, thus isolating the various components. Set the multimeter on the kilo ohms scale. Before testing any part of the CDI unit, bridge the two terminals to be tested with a length of insulated wire, to discharge any residual energy in the unit.

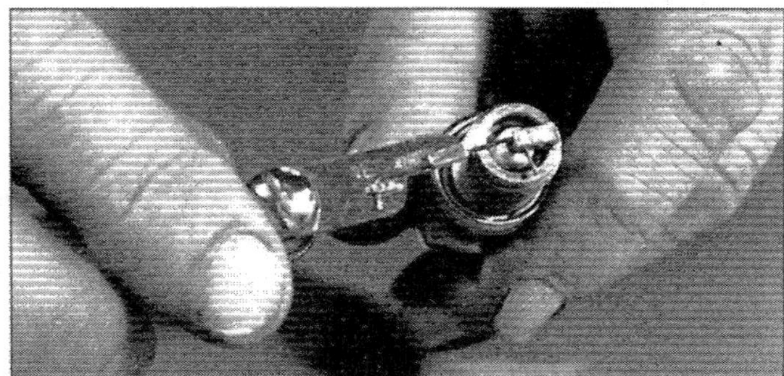

Electrode gap check - use a wire type gauge for best results

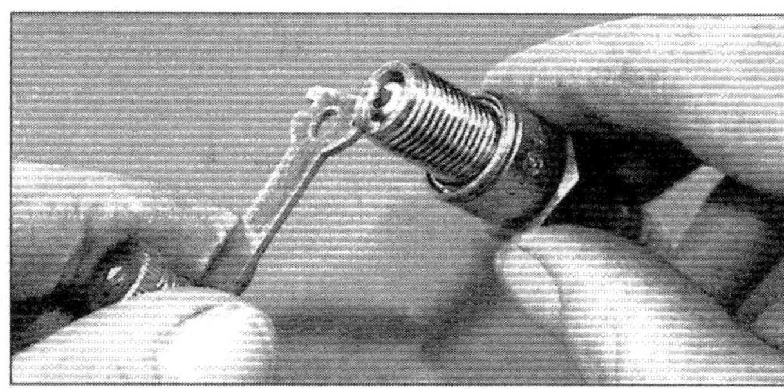

Electrode gap adjustment - bend the side electrode using the correct tool

Normal condition - A brown, tan or grey firing end indicates that the engine is in good condition and that the plug type is correct

Ash deposits - Light brown deposits encrusted on the electrodes and insulator, leading to misfire and hesitation. Caused by excessive amounts of oil in the combustion chamber or poor quality fuel/oil

Carbon fouling - Dry, black sooty deposits leading to misfire and weak spark. Caused by an over-rich fuel/air mixture, faulty choke operation or blocked air filter

Oil fouling - Wet oily deposits leading to misfire and weak spark. Caused by oil leakage past piston rings or valve guides (4-stroke engine), or excess lubricant (2-stroke engine)

Overheating - A blistered white insulator and glazed electrodes. Caused by ignition system fault, incorrect fuel, or cooling system fault

Worn plug - Worn electrodes will cause poor starting in damp or cold weather and will also waste fuel

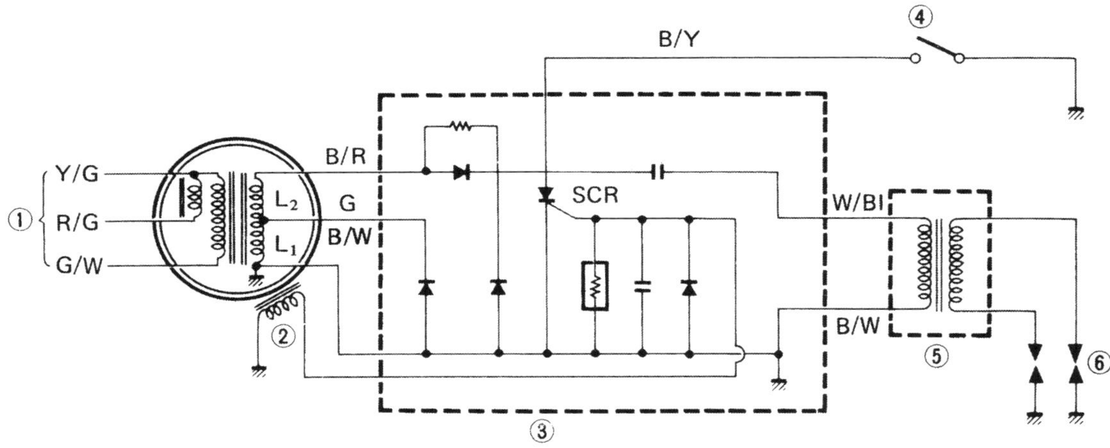

Fig. 3.2 Ignition circuit – GT250 X7 (other models similar)

1 Charging coils
2 Pulser coil
3 CDI unit
4 Ignition switch
5 Ignition coil
6 Spark plug

4 Test procedure – X7 model

a) CDI unit

Negative (-) probe connections	Positive (+) probe connections: B/R	G	B/Y	R/W	W/BL	B/W
B/R	–	*INF*	*CON*	*INF*	*CAP*	*INF*
G	*INF*	–	*INF*	*INF*	*INF*	*INF*
B/Y	*INF*	*INF*	–	*INF*	*CAP*	*INF*
R/W	*INF*	*CON*	*CON*	–	*CAP*	*CON*
W/BL	*INF*	*INF*	*CAP*	*INF*	–	*INF*
B/W	*INF*	*CON*	*CON*	*CON*	*CAP*	–

Key:

B/R	*Black with red tracer*
G	*Green*
B/Y	*Black with yellow tracer*
R/W	*Red with white tracer*
W/BL	*White with blue tracer*
B/W	*Black with white tracer*
INF	*Indicates infinite resistance reading*
CON	*Indicates continuity, or no resistance*
CAP	*Indicates capacitor – needle should deflect momentarily, then settle back to infinity.*

b) Ignition coil

Connect one probe lead to the Black/white terminal, and the other probe to the White/blue terminal. A reading of about 1 ohm should be given. Connect the multimeter between the two high tension leads. A reading of about 10 kilo ohms should be obtained.

c) Exciter coil

Black/red to green terminals: about 30 ohms
Green to black/white terminals: about 200 ohms

If an open or short circuit is indicated in either test, the exciter coil assembly will need to be renewed.

d) Pulser coil

Measure between the red/white lead and the black/white lead. A reading of about 70 ohms should be obtained.

5 Test procedure – X5 and SB200 models

a) CDI unit

Negative (-) probe connections	Positive (+) probe connections: B/Y	B	B/R	B/W	R/W	W/BL
B/Y	–	*CON*	*INF*	*INF*	*INF*	*INF*
B	*CON*	–	*INF*	*INF*	*INF*	*INF*
B/R	*INF*	*INF*	–	*INF*	*INF*	*INF*
B/W	*INF*	*INF*	*3-5*	–	*0-1*	*INF*
R/W	*INF*	*INF*	*3-5*	*0-1*	–	*INF*
W/BL	*INF*	*INF*	*10-20*	*3-5*	*3.5*	–

Key:

B/Y	*Black with yellow tracer*
B	*Black*
B/R	*Black with red tracer*
B/W	*Black with white tracer*
R/W	*Red with white tracer*
W/BL	*White with blue tracer*
CON	*Indicates continuity, or no resistance*
INF	*Indicates infinite resistance*

Figures indicate reading on K ohms scale.

b) Ignition coil

Measure the resistance between the white/blue and black/white terminals. This should be about 0-1 ohms (0.2 ohms, SB200). Measure between the two high tension leads. A reading of 20-30 K ohms (1.5 K ohms, SB200) should be shown.

c) Exciter and pulser coils

Set the multimeter on the ohms scale, and note the readings obtained at the various positions indicated in the table below.

	B/R	B	Earth
R/W	*300-360*	*310-370*	*30-50*
B/R	–	*5-15*	*260-320*
B	–	–	*270-330*

6 Ignition timing: GT200X5 and SB200

1 The ignition timing on these models is set during initial assembly by the manufacturer. The stator is fixed in relation to the crankcase and the rotor is keyed to the crankshaft. Consequently, there is no need for subsequent adjustment, and no provision is made for this.

7 Ignition timing: adjustment and checking – X7

1 When the ignition timing is set by the factory during assembly, it is arranged so that a mark on the stator corresponds with the centre of the top stator securing screw. If the stator is removed for any reason, it is necessary only to check that this setting is duplicated.
2 The timing may be checked more accurately using a stroboscopic timing lamp. Connect the lamp to the sparking plug on high tension lead according to the manufacturer's instructions. Start the engine, and allow it to run at about 6000 rpm. If the lamp is directed at the flywheel rotor periphery, the marks on the crankcase and rotor should appear to be in alignment. If this is not the case, small adjustments can be made by moving the stator.

7.2 Timing can be checked by stroboscope – X7 only

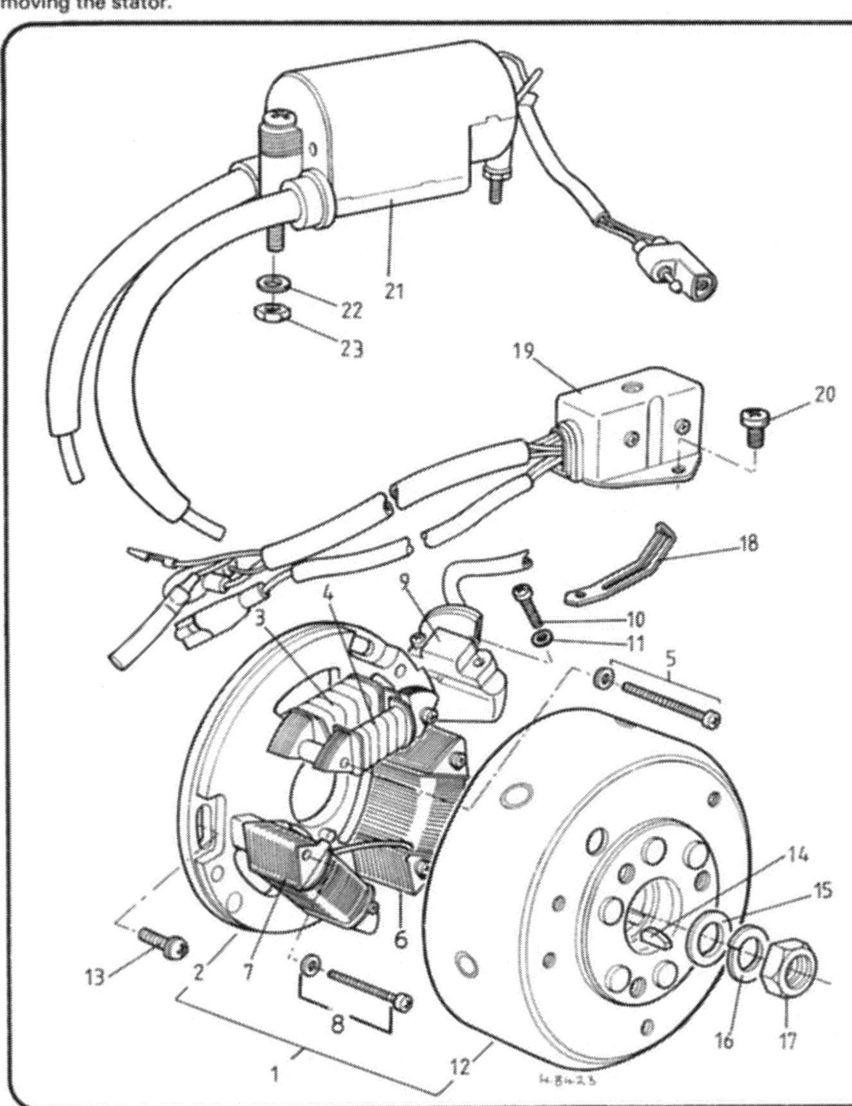

Fig. 3.3 Flywheel generator – X7 (other models similar)

1 Magneto assembly
2 Stator assembly
3 Primary coil A
4 Primary coil B
5 Screw/spring washer – 2 off
6 Charging coil A
7 Charging coil B
8 Screw/spring washer – 4 off
9 Pulser coil
10 Screw – 2 off
11 Spring washer – 2 off
12 Flywheel
13 Screw – 3 off
14 Woodruff key
15 Washer
16 Spring washer
17 Nut
18 Lead clip – 2 off
19 C.D.I unit assembly
20 Screw – 2 off
21 Ignition coil assembly
22 Washer
23 Nut – 2 off

8 Sparking plugs: checking and resetting the gaps

1 Two NGK or Nippon-Denso sparking plugs are fitted to the X7, X5 and SB200 models as standard. The plug types are given in the specifications at the beginning of the Chapter. Certain operating conditions may indicate a change in sparking plug grade, but generally the type recommended by the manufacturer gives the best all round service.
2 Check the gap of the plug points every monthly or 1000 mile service. To reset the gap, bend the outer electrode to bring it closer to, or further away from the central electrode until a 0.7 mm (0.028 in) feeler gauge can be inserted. Never bend the centre electrode or the insulator will crack, causing engine damage if the particles fall into the cylinder whilst the engine is running.
3 With some experience, the condition of the sparking plug electrodes and insulator can be used as a reliable guide to engine operating conditions. See the accompanying series of colour photographs.
4 Always carry spare sparking plugs of the recommended grade. In the rare event of plug failure, this will enable the engine to be restarted.
5 Beware of over-tightening the sparking plugs, otherwise there is risk of stripping the threads from the aluminium alloy cylinder heads. The plugs should be sufficiently tight to seat firmly on their copper sealing washers, and no more. Use a spanner which is a good fit to prevent the spanner from slipping and breaking the insulator.
6 If the threads in the cylinder head strip as a result of over-tightening the sparking plugs, it is possible to reclaim the head by the use of a Helicoil thread insert. This is a cheap and convenient method of replacing the threads; most motorcycle dealers operate a service of this nature at an economic price.
7 Make sure the plug insulating caps are a good fit and have their rubber seals. They should also be kept clean to prevent tracking. These caps contain the suppressors that eliminate both radio and TV interference.

9 Fault diagnosis: ignition system

Symptom	Cause	Remedy
Engine will not start	Short circuit in wiring system	Check wiring to locate source of fault.
	Fouled plug(s)	Remove and clean.
Engine misfires and eventually stops	Whiskered sparking plugs	Check oil pump setting. If correct replace plugs, using a hotter grade
	Oiled sparking plugs	Check oil pump setting. If correct replace plugs, using a colder grade
Engine lacks response and overheats	Incorrect ignition timing (X7)	Check and reset.
Engine 'fades' when under heavy load	Pre-ignition	Replace plugs, using only recommended grades. Check carburettor mountings for air leaks. Check ignition timing (GT250 X7).

Chapter 4 Frame and forks

Contents

Specifications

Frame

Type	Welded tubular steel

Front forks

Type	Hydraulically damped, telescopic
Oil capacity at overhaul* (per leg):	
GT250 X7	150 cc (5.07/5.28 US/Imp oz)
GT200 X5	140 cc (4.73/4.93 US/Imp oz)
SB200	188 cc (6.35/6.62 US/Imp oz)
Oil grade	SAE 10W/20 Motor oil or ATF

Rear suspension

Type	Swinging arm, supported on two oil-damped suspension units

* No fork leg oil capacities are available other than those given. It is suggested that the quantity per leg be reduced by about 10 cc when refilling forks which have been drained only and not dismantled.

1 General description

The Suzuki GT250 X7, GT200 X5 and SB200 each use conventional welded tubular steel frames of similar design.

The front wheel is supported by telescopic front forks, which incorporate oil-filled damper units. Rear suspension is by means of a pivoted rear fork, or swinging arm, controlled by adjustable, hydraulically damped rear suspension units.

2 Front forks: removal and replacement

1 It is unlikely that the front forks will need to be removed from the frame together with the fork yoke and steering stem. Although feasible, this method is both unwieldy and time consuming. It is far easier to remove the individual fork legs, then the steering head assembly, where necessary, as described in Section 3 of this Chapter.

2 Start by placing the machine securely on its centre stand. If desired, wooden blocks can be placed beneath the stand to raise the front wheel well clear of the ground, making removal of the wheel easier. Remove the front wheel as described in Section 4.2 or 9.1 of Chapter 5, and place it to one side.

3 On X5 and X7 models, remove the brake caliper securing bolts to allow the caliper to be tied clear of the forks. Do not allow the caliper to hang from the hydraulic hose. On all models, the front mudguard should be removed after releasing its securing nuts or bolts.

4 On X5 and X7 machines, slacken the pinch bolts on the upper and lower yokes to allow the fork legs to be pulled down and clear of the frame. On SB200 models, slacken the lower yoke pinch bolts and remove the large fork top bolts. The fork legs can now be withdrawn downwards. If they prove reluctant to move, temporarily refit the fork top bolts about half way home, and tap the bolt head with a soft-faced mallet to dislodge the fork leg. The shrouds used on SB200 models can be left in position unless the fork yokes are to be removed.

5 Reassembly is tackled in the reverse order of that given for dismantling. On SB200 models, some difficulty may be encountered when attempting to pass the fork stanchions up through the shrouds, as the stanchion may tend to drop back into the lower leg. This problem can be overcome by using a length of wooden dowel sharpened to a taper at one end. The taper can be 'screwed' into the top bolt thread, and used as a pilot to draw the stanchion into position. Alternatively, a suitable long bolt can be used in a similar fashion. When the stanchion appears to be firmly in place, fit the top bolt to draw it up against the top yoke, then tighten the lower yoke pinch bolt.

6 On X5 and X7 machines, feed the fork legs into position, ensuring that the tops of each stanchion are level. Tighten the upper and lower pinch bolts. On all models, if the steering head assembly has been dismantled, it is important to ensure that the forks are correctly aligned when refitted. To this end, assemble the forks loosely, refit the front wheel, then bounce the forks a few times so that the various components assume their correct relative positions. Tighten the clamp bolts, working from the wheel spindle upwards.

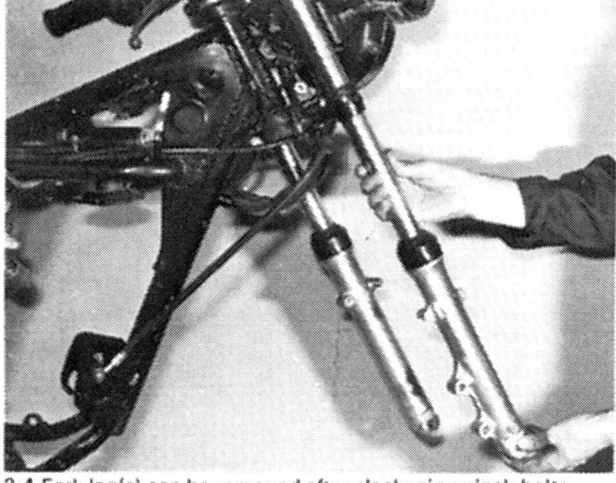
2.4 Fork leg(s) can be removed after slackening pinch bolts

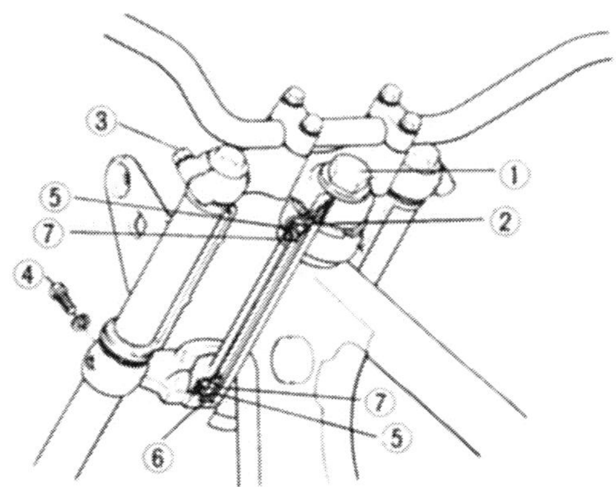

Fig. 4.1 Steering head assembly

1 Steering stem cap bolt
2 Steering stem adjuster nut
3 Top yoke pinch bolt
4 Lower yoke pinch bolt
5 Steel balls
6 Cup
7 Cone

3 Steering head assembly: dismantling and reassembly

1 Commence by following the dismantling procedure detailed in Section 2, paragraphs 1 to 4. Cover the petrol tank with some thick layers of rag, or remove it completely, to prevent damage to the paintwork. Slacken and remove the handlebar clamp bolts. The handlebar assembly and control levers can be laid across the protected tank, clear of the steering head. Release any cable or brake pipe guide clips, so that the handlebar assembly is completely free of the fork yokes.

2 Slacken and remove the steering stem bolt. Pull the upper yoke free from the steering stem, taking great care not to strain the various leads going into the instrument panel. It may prove necessary to tap the yoke very lightly on its underside, to disengage it from the steering stem. Once freed, the upper yoke and instruments can be manoeuvered clear of the steering stem and lower yoke.

3 Before proceeding further, it should be noted that the steering head bearings are of the uncaged ball type, and these will drop free as the steering stem and lower yoke are released. The steel balls will drop free if left unattended, so some provision must be made to catch them. It is a good idea to place some clean cloth under the steering head area to contain any errant balls.

4 Using a C-spanner, slacken and remove the steering stem nut, whilst supporting the lower yoke. With the nut removed, carefully lower the yoke and steering stem, catching the balls from the lower race as they drop free. It will be noted that there are 22 balls in the upper steering head race, and 18 in the lower race.

5 The steering head assembly should be reassembled in the reverse order of that given for dismantling. When fitting the steel balls to their races, they can be held in place with grease. Check that the correct number is fitted to each bearing race. When fitting the steering stem nut, it must be adjusted so that all perceptible free play is taken up, but no more. It is easy to damage the head races by overtightening. When correctly adjusted, it should be possible to move the steering from lock to lock with the lightest pressure on the handlebar end. Final adjustment can be made after reassembly, by slackening the top bolt and adjusting the steering stem nut as required.

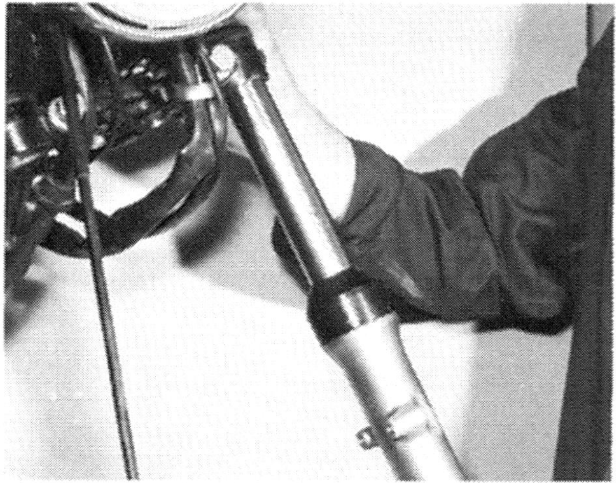
3.1a Remove remaining fork leg before dismantling yokes

3.1b Release all cables, hoses or wiring clips

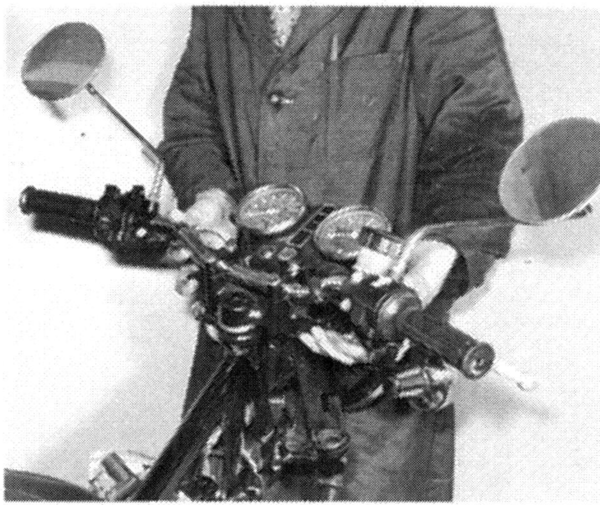
3.2 Handlebar, instruments and top yoke can now be removed

3.4a Unscrew slotted adjuster nut ...

3.4b ... and lift away the dust seal ...

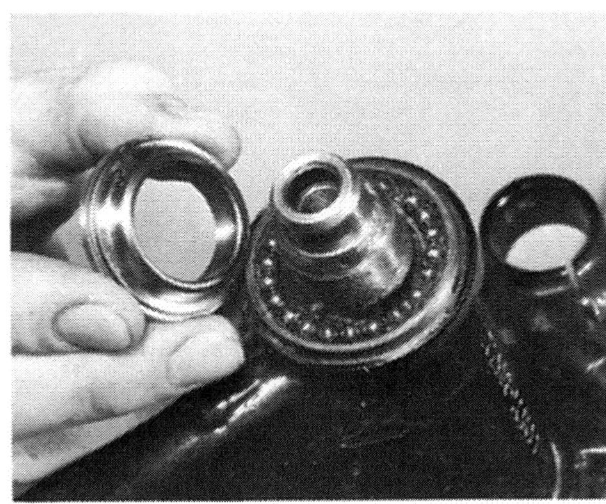
3.4c ... and upper cone

3.5 Balls can be retained with grease

4 Front forks: dismantling and reassembly – SB200

1 With the front fork legs removed, as described in section 2, slide off the upper spring seat, the external spring and the spring shroud. Invert the legs over a suitable drain tray or bowl, and allow the oil to drain. This process can be accelerated by 'pumping' the oil out of the legs by telescoping them. One leg should be tackled at a time, to avoid any possibility of parts becoming interchanged between the two fork legs.
2 Slide off the dust seal and washer to expose the circlip at the top of the lower leg. Release the circlip, and remove the remaining plain washer. The oil seal can be removed after the fork stanchion and damper assembly have been withdrawn.
3 Invert the leg, and remove the damper bolt from the underside of the lower leg. It may prove necessary to apply pressure on the damper assembly whilst the bolt is released. This can be accomplished by passing a long bar down the inside of the stanchion and pushing it against the damper. The stanchion and damper tube assembly can now be displaced for examination. The oil seal should be renewed as a matter of course. Lever out the old seal using a screwdriver or similar.
4 Reassemble the fork leg in the reverse order of the dismantling sequence, noting the following points. When fitting the new oil seal, use a suitably sized socket as a drift to ensure that it is fitted squarely into its recess. Take care not to damage the seal when it is being fitted. Check that the plain washer, circlip and second plain washer are fitted in the correct sequence. Before fitting the stanchion, lubricate the new seal face liberally with oil. The reassembled fork should be topped up with the recommended quantity and grade of damping oil. See the specifications section for details.

5 Front forks: dismantling and reassembly – X5 and X7

1 Remove the fork legs from the frame as detailed in Section 2 of this Chapter. Remove the fork top bolt from the stanchion of each leg, and withdraw the spring (X7) or spacer and spring (X5). Invert each leg over a bowl or drain tray and 'pump' the leg repeatedly to expell the old damping oil. Dismantle one leg at a time to avoid the accidental interchanging of parts.
2 Prise the dust seal off the lower leg and slide it off the end of the stanchion. It is now necessary to remove the damper securing bolt from the base of the lower leg, an operation which usually leads to the damper turning with the bolt. In the absence of the correct holding tool, a wooden or metal rod, ground to a taper can be introduced down the stanchion and pushed against the damper to prevent it from turning. Remove the damper retaining bolt, and draw the stanchion out of the lower leg.
3 The damper assembly can be tipped out of the stanchion for examination. Prise out the snap ring from the top of the lower leg, and remove the plain washer. The seal can be prised out with a screwdriver, and should be renewed as a matter of course.
4 When reassembling the forks, fit the new seals using a large socket as a drift to ensure that they seat squarely. Lubricate the seal faces with engine oil before introducing the stanchion. Fill the forks with the recommended quantity and grade of damping oil (see specifications for details).

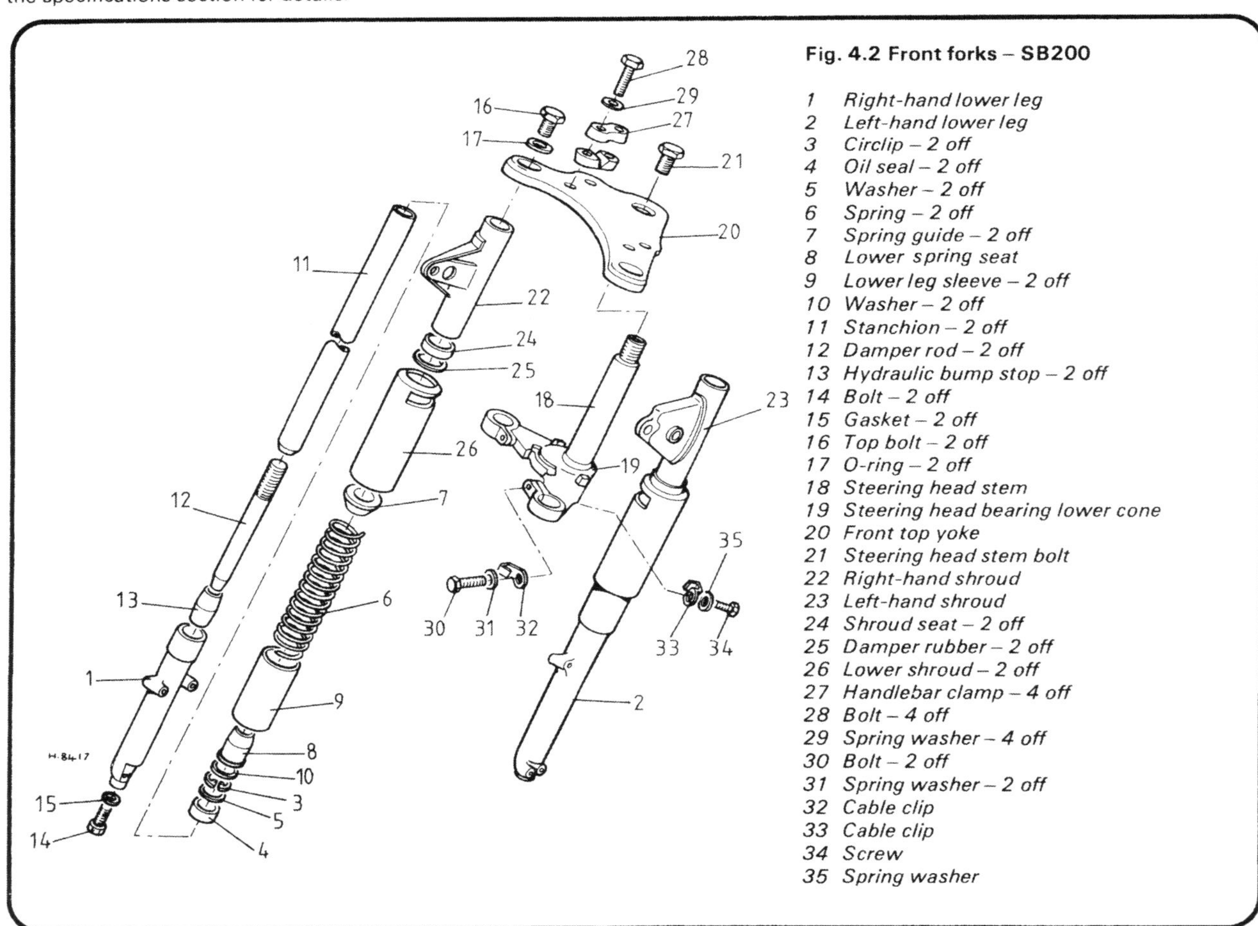

Fig. 4.2 Front forks – SB200

1 Right-hand lower leg
2 Left-hand lower leg
3 Circlip – 2 off
4 Oil seal – 2 off
5 Washer – 2 off
6 Spring – 2 off
7 Spring guide – 2 off
8 Lower spring seat
9 Lower leg sleeve – 2 off
10 Washer – 2 off
11 Stanchion – 2 off
12 Damper rod – 2 off
13 Hydraulic bump stop – 2 off
14 Bolt – 2 off
15 Gasket – 2 off
16 Top bolt – 2 off
17 O-ring – 2 off
18 Steering head stem
19 Steering head bearing lower cone
20 Front top yoke
21 Steering head stem bolt
22 Right-hand shroud
23 Left-hand shroud
24 Shroud seat – 2 off
25 Damper rubber – 2 off
26 Lower shroud – 2 off
27 Handlebar clamp – 4 off
28 Bolt – 4 off
29 Spring washer – 4 off
30 Bolt – 2 off
31 Spring washer – 2 off
32 Cable clip
33 Cable clip
34 Screw
35 Spring washer

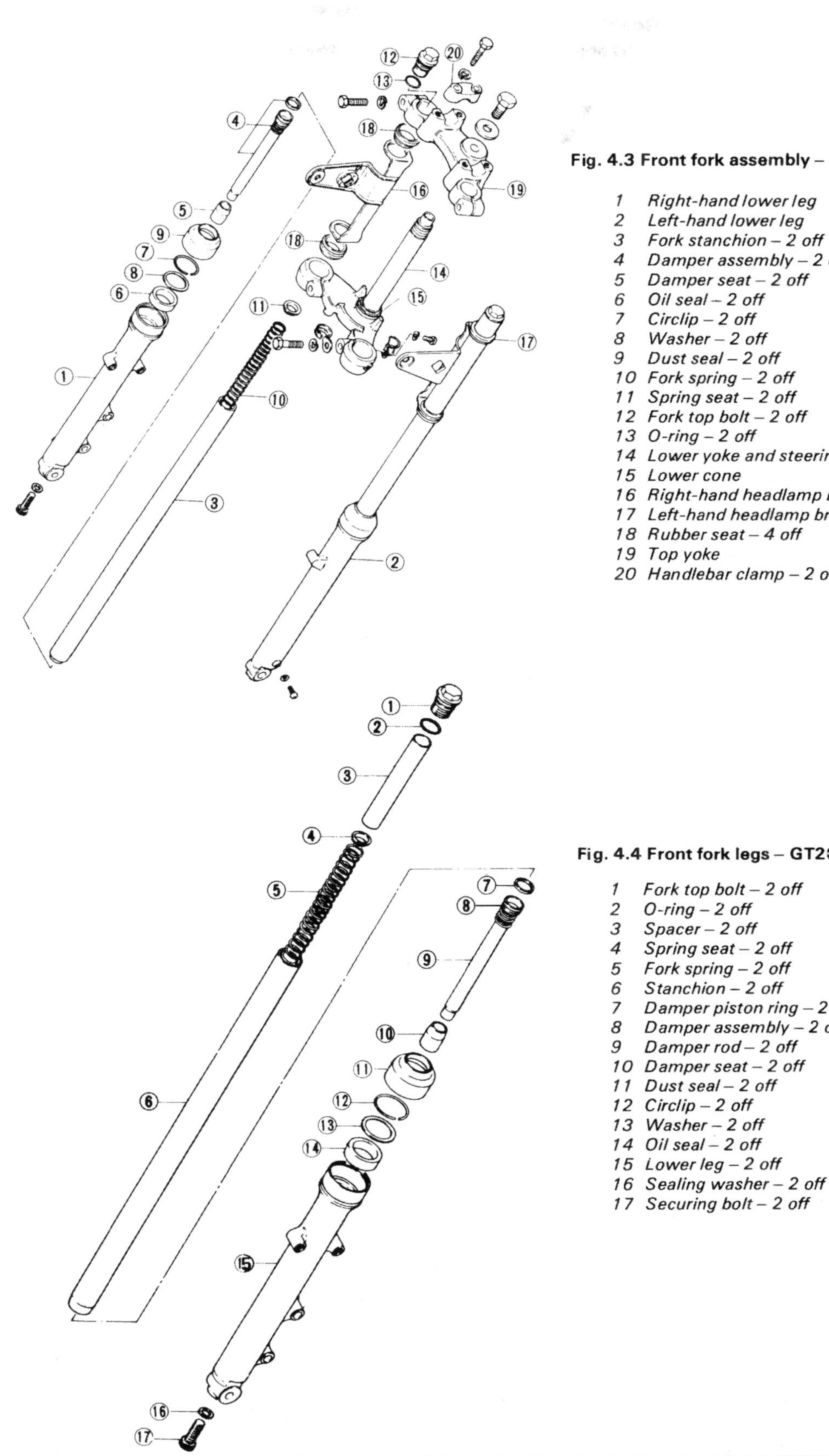

Fig. 4.3 Front fork assembly – GT250 X7

1 *Right-hand lower leg*
2 *Left-hand lower leg*
3 *Fork stanchion – 2 off*
4 *Damper assembly – 2 off*
5 *Damper seat – 2 off*
6 *Oil seal – 2 off*
7 *Circlip – 2 off*
8 *Washer – 2 off*
9 *Dust seal – 2 off*
10 *Fork spring – 2 off*
11 *Spring seat – 2 off*
12 *Fork top bolt – 2 off*
13 *O-ring – 2 off*
14 *Lower yoke and steering stem*
15 *Lower cone*
16 *Right-hand headlamp bracket*
17 *Left-hand headlamp bracket*
18 *Rubber seat – 4 off*
19 *Top yoke*
20 *Handlebar clamp – 2 off*

Fig. 4.4 Front fork legs – GT200 X5

1 *Fork top bolt – 2 off*
2 *O-ring – 2 off*
3 *Spacer – 2 off*
4 *Spring seat – 2 off*
5 *Fork spring – 2 off*
6 *Stanchion – 2 off*
7 *Damper piston ring – 2 off*
8 *Damper assembly – 2 off*
9 *Damper rod – 2 off*
10 *Damper seat – 2 off*
11 *Dust seal – 2 off*
12 *Circlip – 2 off*
13 *Washer – 2 off*
14 *Oil seal – 2 off*
15 *Lower leg – 2 off*
16 *Sealing washer – 2 off*
17 *Securing bolt – 2 off*

5.1 Release top bolt and spacer, where fitted

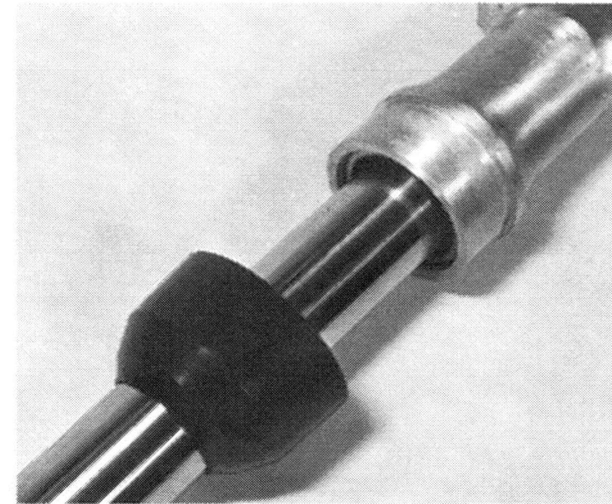

5.2a Dust seal can be slid off the stanchion

5.2b Cut slot in length of tubing and fit plain washer

5.2c Improvised holding tool holds damper thus

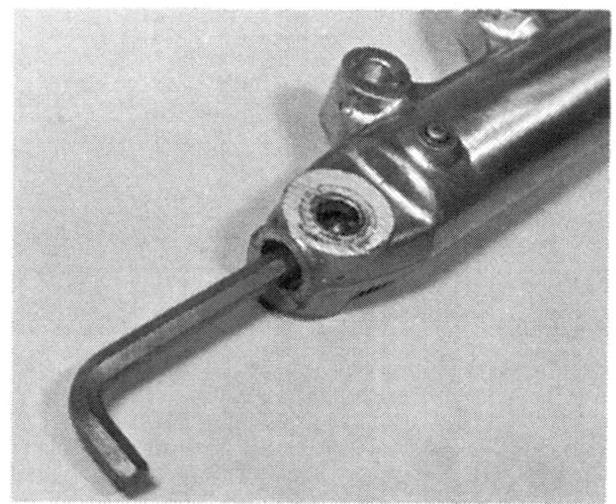

5.2d Slacken damper retaining bolt ...

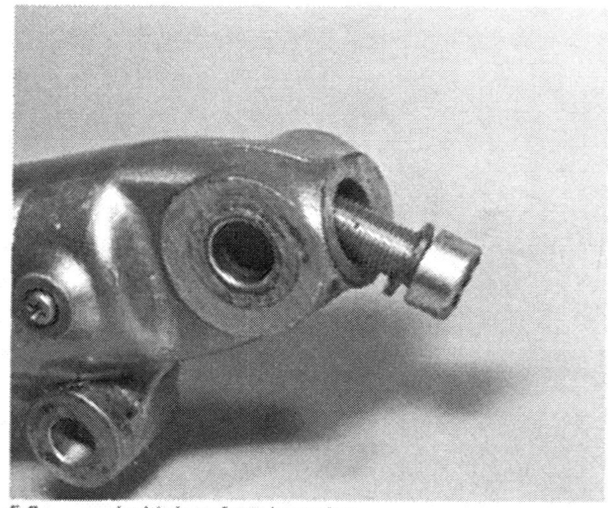

5.2e ... and withdraw from lower leg

5.3a Check sliding surfaces for scoring

5.3b Remove snap ring and plain washer ...

5.3c ... to allow seal removal

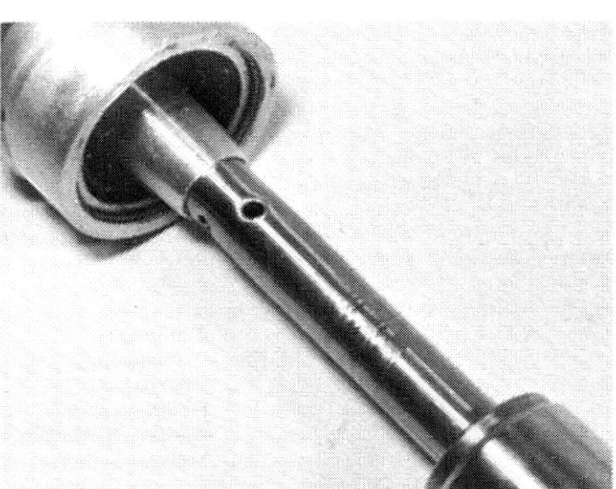
5.4a Grease seal prior to reassembly

5.4b Tightest coils should be uppermost

6 Front forks: examination and renovation

1 The parts most liable to wear over an extended period of service are the internal surfaces of the lower leg and the outer surfaces of the fork stanchion or tube. If there is excessive play between these two parts, they must be renewed as a complete unit. Check the fork tube for scoring over the length which enters the oil seal. Bad scoring here will damage the oil seal and lead to fluid leakage.

2 It is advisable to renew the oil seals when the forks are dismantled, even if they appear to be in good condition. This will save a strip down of the forks at a later date if oil leakage occurs. The oil seal in the top of each lower fork leg is retained by an internal C-ring which can be prised out of position with a small screwdriver. Check that the dust excluder rubbers are not split or worn where they bear on the fork tube. A worn excluder will allow the ingress of dust and water which will damage the oil seal and eventually cause wear of the fork tube.

3 It is not generally possible to straighten forks which have been badly damaged in an accident, particularly when the correct jigs are not available. It is always best to err on the side of safety and fit new ones, especially since there is no easy means to detect whether the forks have been over stressed or

metal fatigued. Fork stanchions (tubes) can be checked, after removal from the lower legs by rolling them on a dead flat surface. Any misalignment will be immediately obvious.

4 The fork springs will take a permanent set after considerable usage and will need renewal if the fork action becomes spongy. The service limit for the total free length of each spring is as follows.

GT250 X7	*416 mm (16.4 in)*
GT200 X5	*315 mm (12.4 in)*
SB200	*167.8 mm (6.6 in)*

5 When renewing springs it should be remembered that they must always be renewed as a pair and not singly.

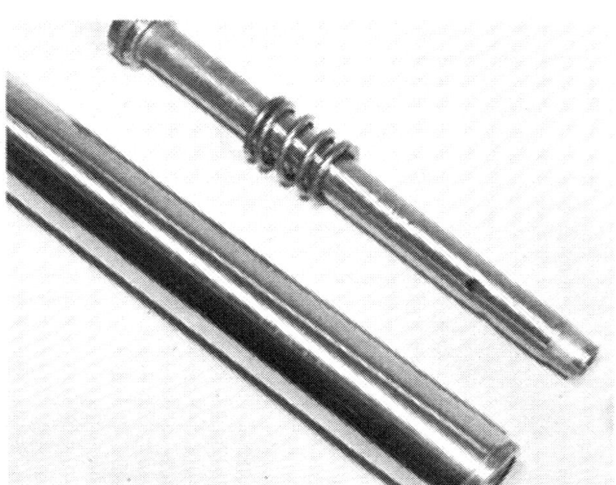
6.1 Renew stanchion if damaged by scoring

7 Steering head bearings: examination and renovation

1 Before commencing reassembly of the forks, examine the steering head races. The ball bearing tracks of the respective cup and cone bearings should be polished and free from indentations, cracks or pitting. If signs of wear are evident, the cups and cones must be renewed. In order for the straight line steering on any motorcycle to be consistently good, the steering head bearings must be absolutely perfect. Even the smallest amount of wear on the cups and cones may cause steering wobble at high speeds and judder during heavy front wheel braking. The cups and cones are an interference fit on their respective seatings and can be tapped from position with a suitable drift.

2 Ball bearings are relatively cheap. If the originals are marked or discoloured they must be renewed. To hold the steel balls in place during reassembly of the fork yokes, pack the bearings with grease. The upper race is fitted with twenty-two steel balls and the lower race eighteen balls. Although each race has room for an extra steel ball it must not be fitted. The gap allows the bearings to work correctly, stopping them skidding and accelerating the rate of wear.

8 Steering lock

1 The steering lock is mounted forward of the steering head lug where it is secured by a single bolt. When the steering is turned fully to the right or left and the lock is operated the mounting bolt is inaccessible and therefore unauthorised removal is prevented.

2 If the lock malfunctions repair is impracticable and a new lock must be fitted. It follows that new keys must be acquired at the same time.

9 Frame: examination and renovation

1 The frame is unlikely to require attention unless accident damage has occurred. In some cases, replacement of the frame

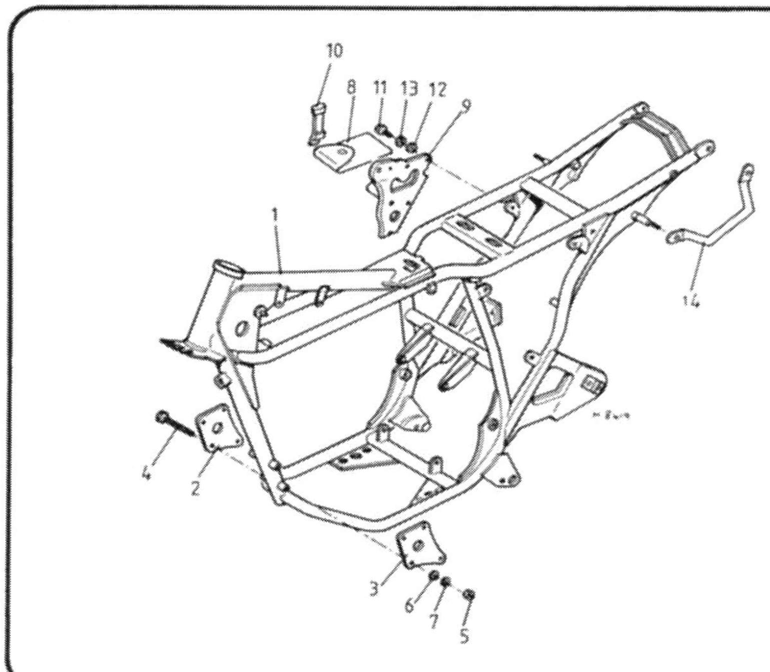

Fig. 4.5 Frame – SB200 (other models similar)

1 Frame
2 Right-hand engine plate
3 Left-hand engine plate
4 Bolt – 2 off
5 Nut – 2 off
6 Washer – 2 off
7 Locking washer – 2 off
8 Tool kit
9 Bracket
10 Strap
11 Bolt – 3 off
12 Washer
13 Locking washer – 3 off
14 Lifting handle

is the only satisfactory course of action if it is badly out of alignment. Only a few frame repair specialists have the jigs and mandrels necessary for resetting the frame to the required standard of accuracy and even then there is no easy means of assessing to what extent the frame may have been overstressed.

2 After the machine has covered a considerable mileage, it is advisable to examine the frame closely for signs of cracking or splitting at the welded joints. Rust can also cause weakness at these joints. Minor damage can be repaired by welding or brazing, depending on the extent and nature of the damage.

3 Remembering that a frame which is out of alignment will cause handling problems and may even promote 'speed wobbles'. If misalignment is suspected, as the result of an accident, it will be necessary to strip the machine completely so that the frame can be checked and, if necessary, renewed.

10 Swinging arm rear fork: examination and renovation

1 The rear fork assembly pivots on shouldered bushes pressed into the cross-member at the end of each fork arm. (X5 and SB200) or needle roller bearings (X7), which are supported by a pivot shaft running through the frame tubes. Worn swinging arm pivot bearings will give imprecise handling, with a tendency for the rear end of the machine to twitch or hop. The play can be detected by placing the machine on its centre stand and with the rear wheel clear of the ground, pulling and pushing on the fork ends in a horizontal direction. Any play will be magnified by the leverage effect. In the UK, excess play will cause the machine to fail the DOT test. It is quite easy to renovate the swinging arm when wear necessitates attention.

2 To remove the rear swinging arm fork, first position the machine on the centre stand so that it rests firmly and securely. Remove the final drive chain by detaching the master link, and then unscrew the two bolts that hold the chain guard. On GT250 X7 models not fitted with a chain link the chain should be detached from the rear wheel sprocket after removal of the wheel spindle.

3 Detach the brake torque arm from the lug on the brake plate and from the lug on the frame. In both cases the torque arm bolt retaining nuts are secured by split pins. Unscrew the adjuster nut on the brake operating arm and pull the rod through the trunnion on the brake arm. Replace the adjuster nut to avoid the loss of the brake rod spring.

4 Remove the split pin from the end of the wheel spindle and remove the castellated nut. The wheel spindle can now be pulled out to the right. Knock the wheel spacer out of position from between the brake plate and the fork leg. The rear wheel can now be lifted rearwards, out of the frame.

5 Remove the lower nut from each rear suspension unit and swing the units back, out of the way.

6 Remove the locknut from the end of the pivot shaft, which can then be tapped out from the left-hand side. Working the swinging arm fork up and down will aid removal of the shaft. The swinging arm fork is now free to be pulled from position between the two frame lugs.

7 Remove the dust excluder caps from each end of the fork crossmember. Note the presence of the O-rings in the caps. Push out the two inner bushes.

8 Wash the inner and outer bushes carefully in petrol or another solvent. Do not remove the outer bushes from position in the fork crossmember unless they need renewal as they are made of a brittle material that will probably fracture while being drifted out. When drifting in new bushes, ensure that they enter their housings squarely. Use a soft wood or hard rubber pad between the bush and drive, to prevent chipping. A better alternative to this method, particularly in the case of the X7's needle roller bearings, is to use a drawbolt arrangement to pull the bearings into position. Check the pivot shaft for straightness by rolling it on the edge of a dead flat surface. If the shaft is bent it must be renewed or straightened.

9 Reassemble the swinging arm fork by reversing the dismantling procedure. Grease the pivot shaft and bearings liberally before reassembly and check that the sealing rings in the dust caps are in good condition. (X5 and SB200 only).

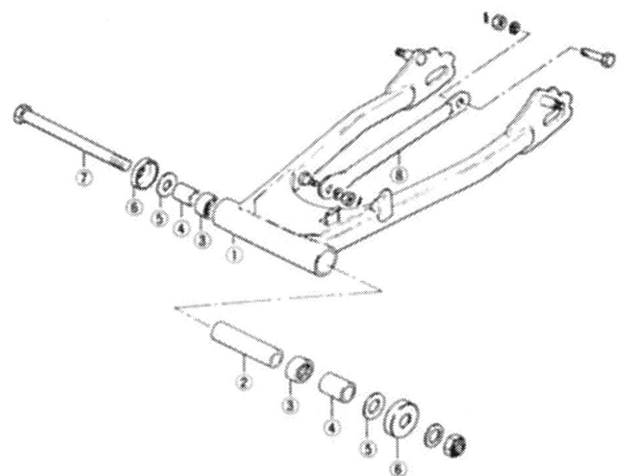

Fig. 4.6 Swinging arm assembly

1 *Swinging arm fork*
2 *Bearing spacer*
3 *Bearing (X7) or bush (X5 and SB200)*
4 *Inner bush*
5 *Thrust washer*
6 *Dust cover*
7 *Pivot shaft*
8 *Rear brake torque arm*

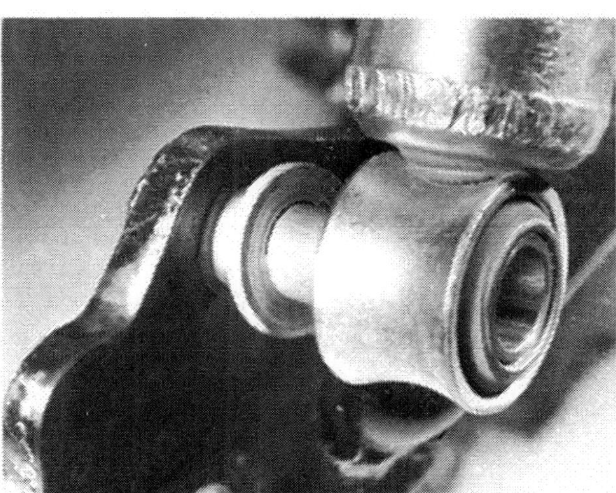

10.5 Rear suspension units are secured by nuts

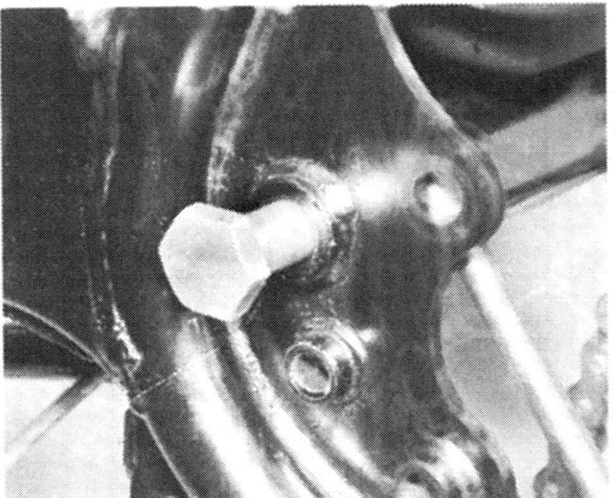
10.6a Withdraw pivot shaft ...

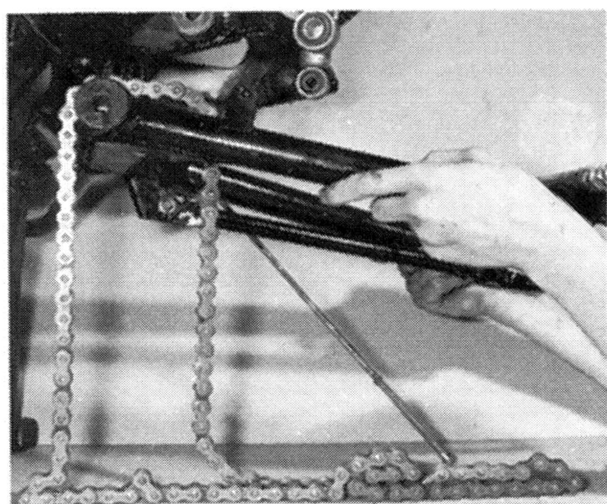
10.6b ... and remove swinging arm

10.7 Remove dust caps

10.8 X7 has needle roller bearings

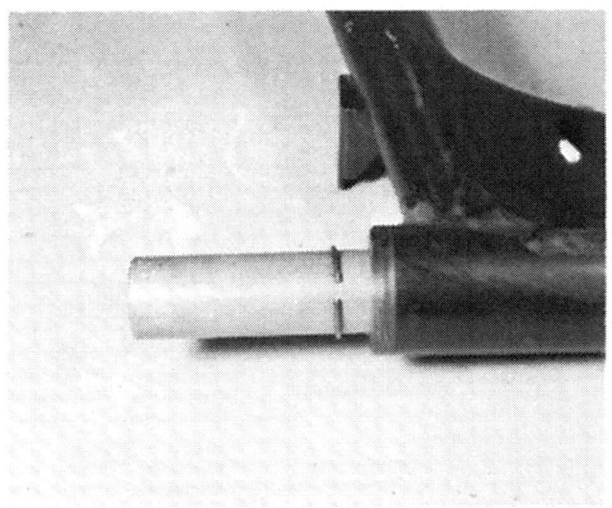
10.9a Spacer is fitted between the bearings

10.9b Grease inner bushes prior to installation

11 Rear suspension units: examination

1 The rear suspension units fitted to the Suzuki X7, and X5 models are of the normal, hydraulically damped type, adjustable to give five different spring settings. A C-spanner included in the tool kit should be used to turn the lower spring seat and so alter its position on the adjustment projection. When the spring seat is turned so that the effective length of the spring is shortened, the suspension will become heavier. The units fitted to the SB200 are not adjustable.
2 If a suspension unit leaks, or if the damping efficiency is reduced in any other way, the two units must be renewed as a pair. For precise roadholding, it is imperative that both units react to movement in the same way. It follows that the units must always be set at the same spring loading.

12 Centre stand: examination and maintenance

1 The centre stand is an important but largely neglected feature of most motorcycles. It is important to check the stand for wear or damage from time to time, as failure of the stand can result in costly repair bills. Check that the stand mounting bolts are secure and in good condition, and that they are kept adequately lubricated.
2 Check that the return spring is in good condition. A broken or weak spring may cause the stand to fall whilst the machine is being ridden, and catch in some obstacle, unseating the rider.

13 Prop stand: examination

1 The prop stand is attached to a lug welded to the left-hand lower frame tube. An extension spring anchored to the frame ensures that the stand is retracted when the weight of the machine is taken off the stand.
2 Check that the pivot bolt is secured and that the extension spring is in good condition and not overstretched. An accident is almost certain if the stand extends whilst the machine is on the move.

14 Footrests: examination and renovation

1 The front footrests take the form of an assembly bolted to the underside of the machine (separate assemblies on X7 model). It is possible that this assembly will become bent if the machine is dropped, as the footrests are not of the folding type. It is preferable to renew the damaged parts, but if necessary, they can be bent straight by clamping them in a vice and heating to a dull red with a blow lamp whilst the appropriate pressure is applied. Do not attempt to straighten the footrests while they are attached to the frame.
2 If heat is applied to the main footrest piece during any straightening operation it follows that the footrest rubber must be removed temporarily.
3 The rear footrests are of the folding type and are unlikely to require frequent attention. Little can be done to repair them, and in the event of extreme wear or damage, the affected parts should be renewed. If the footrests are damaged in an accident, it is possible to dismantle the assembly into its component parts. Detach each footrest from the mounting plate by withdrawing the split pin and pulling out the clevis pin.

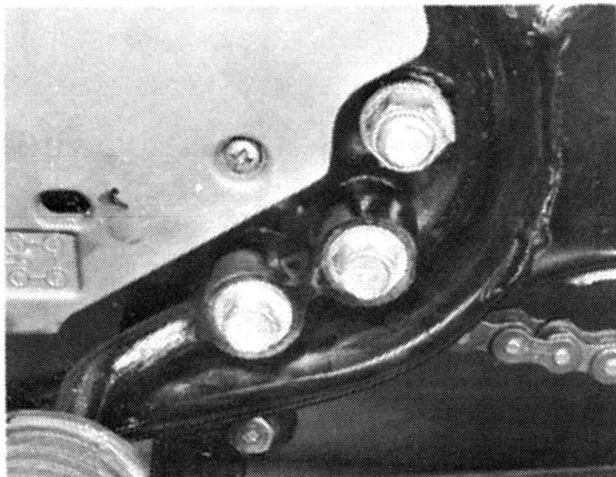
14.1 Footrests bolt to frame

15 Rear brake pedal: examination and renovation

1 The rear brake pedal pivots on a pin projecting from the frame, and is secured by a washer and split pin. If damaged, the pedal can be removed after the split pin has been extracted.
2 The pedal should be straightened by adopting the same method as recommended for bent footrests. It should be borne in mind that heating the pedal will almost certainly destroy the chrome plate with which the component is finished, therefore if the cosmetic appearance of the machine is important the part should be renewed.
3 The rear brake pedal is returned to its normal position by an extension spring. This should be checked to ensure that it is not stretched, and pulls the brake off cleanly.

16 Speedometer head and tachometer head: removal and replacement

1 The speedometer head and the tachometer head (where fitted) may be removed from the machine individually by removing the instrument console cover and releasing the two nuts or screws holding each instrument, or together by detaching the complete console and then separating the components, after removal. If the former method is adopted, the bulb holders should be removed from the base of the instruments as they are lifted from position.
2 The drive cables must be detached first in either case. Each is secured by a knurled ring. Apart from defects in either the drive or drive cables, a speedometer or tachometer which malfunctions is difficult to repair. Fit a replacement or alternatively entrust the repair to a competent instrument repair specialist.
3 Remember that a speedometer in correct working order is a statutory requirement in the UK. Apart from this legal necessity, reference to the odometer readings is the most satisfactory means of keeping pace with the maintenance schedules.

17 Speedometer and tachometer drive cables: examination and maintenance

1 It is advisable to detach the drive cable(s) from time to time in order to check whether the outer coverings are damaged or compressed at any point along their run. Jerky or sluggish movements can often be traced to a damaged drive cable.
2 It is not practicable to effect a satisfactory repair to a damaged or broken drive cable, and in this event the complete cable must be renewed.

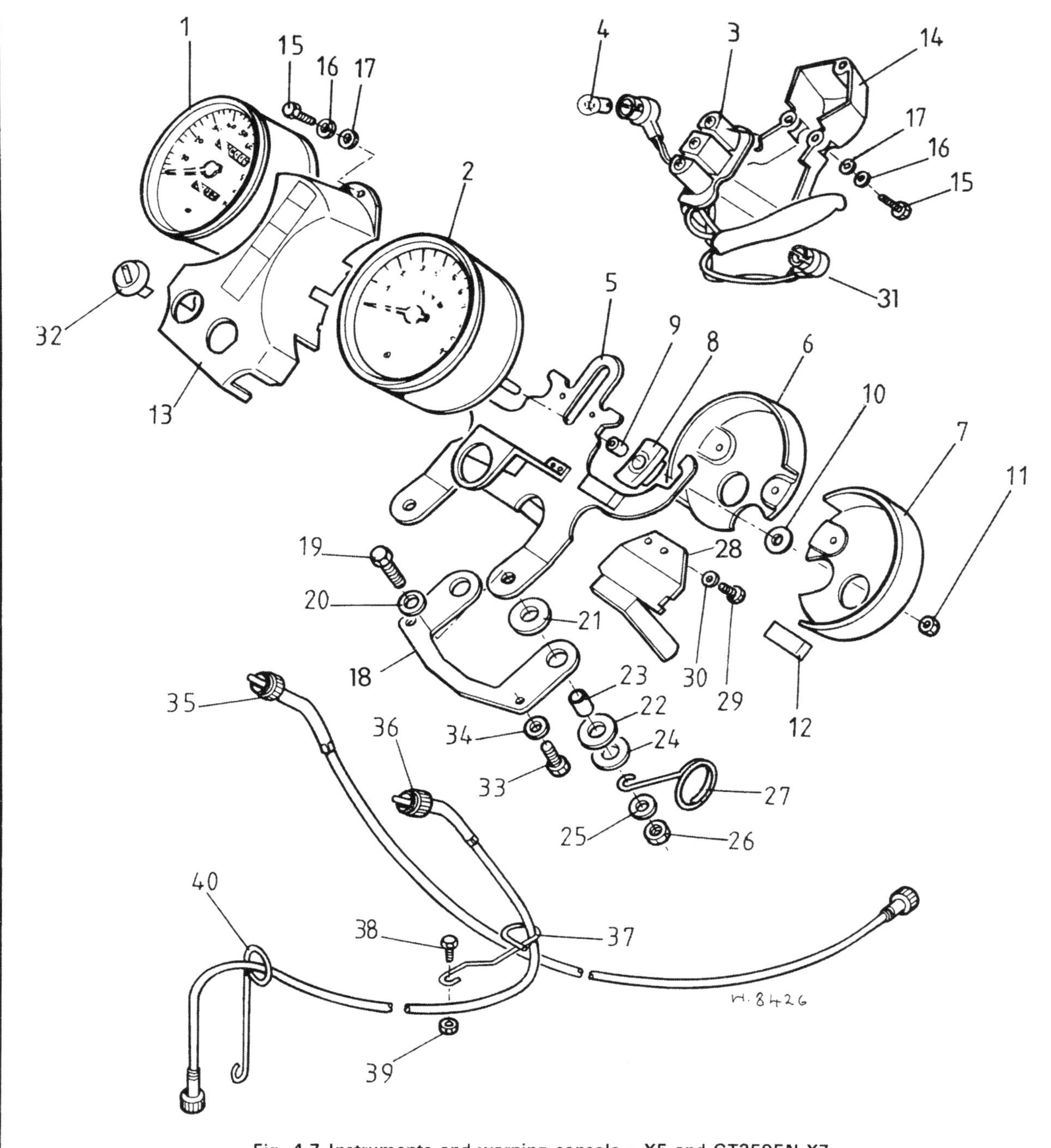

Fig. 4.7 Instruments and warning console – X5 and GT250EN X7

1 Speedometer assembly
2 Tachometer assembly
3 Bulbholder
4 Bulb – 5 off
5 Instrument mounting bracket
6 Speedometer lower cover
7 Tachometer lower cover
8 Grommet – 4 off
9 Spacer – 4 off
10 Washer – 4 off
11 Nut – 4 off
12 Damper
13 Warning panel
14 Fixing bracket
15 Screw – 3 off
16 Spring washer – 3 off
17 Washer – 3 off
18 Cable bracket
19 Bolt – 2 off
20 Washer – 2 off
21 Grommet – 2 off
22 Grommet – 2 off
23 Spacer – 2 off
24 Washer – 2 off
25 Spring washer – 2 off
26 Nut – 2 off
27 Throttle cable guide
28 Bracket
29 Screw
30 Spring washer
31 Bulbholder
32 Logos
33 Bolt – 2 off
34 Washer – 2 off
35 Speedometer cable assembly
36 Tachometer cable assembly
37 Tachometer cable guide
38 Bolt
39 Nut
40 Tachometer cable guide

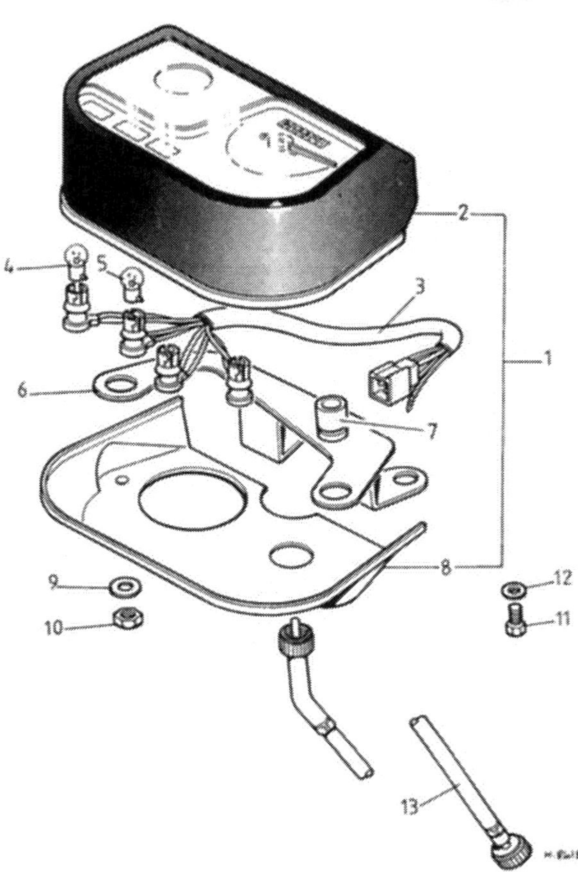

Fig. 4.8 Speedometer assembly – SB200

1 Speedometer assembly
2 Speedometer
3 Bulb holders and cable
4 Bulb – 3 off
5 Bulb
6 Warning instruments mounting bracket
7 Grommet
8 Speedometer lower cover
9 Washer
10 Nut
11 Bolt
12 Washer
13 Speedometer cable assembly

18 Speedometer and tachometer drives: location and examination

1 In the case of the disc front brake machines, the speedometer drive gearbox is fitted on the left-hand side of the front wheel hub. On drum front brake machines the gearbox is an integral part of the brake plate and is driven internally from the front hub. In both cases the drive rarely gives trouble provided it is kept properly lubricated. Lubrication should take place whenever the front wheel is removed for wheel bearing inspection or replacement.

2 The tachometer drive is taken from the primary drive cover via a gear on the end of the oil pump driven pinion (crankshaft, on SB 200), and then through a flexible cable to the tachometer head. It is unlikely that the internal drive will give trouble during the normal service life of the machine, particularly since it is fully enclosed and effectively lubricated.

16.1 Instruments are held by nuts (arrowed)

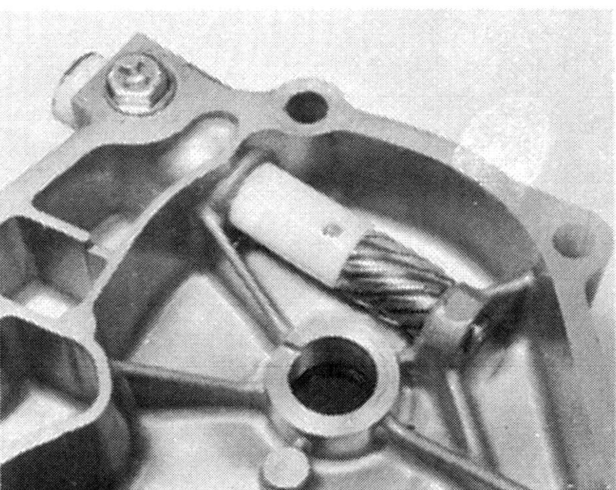

18.2 Tachometer drive rarely causes problems

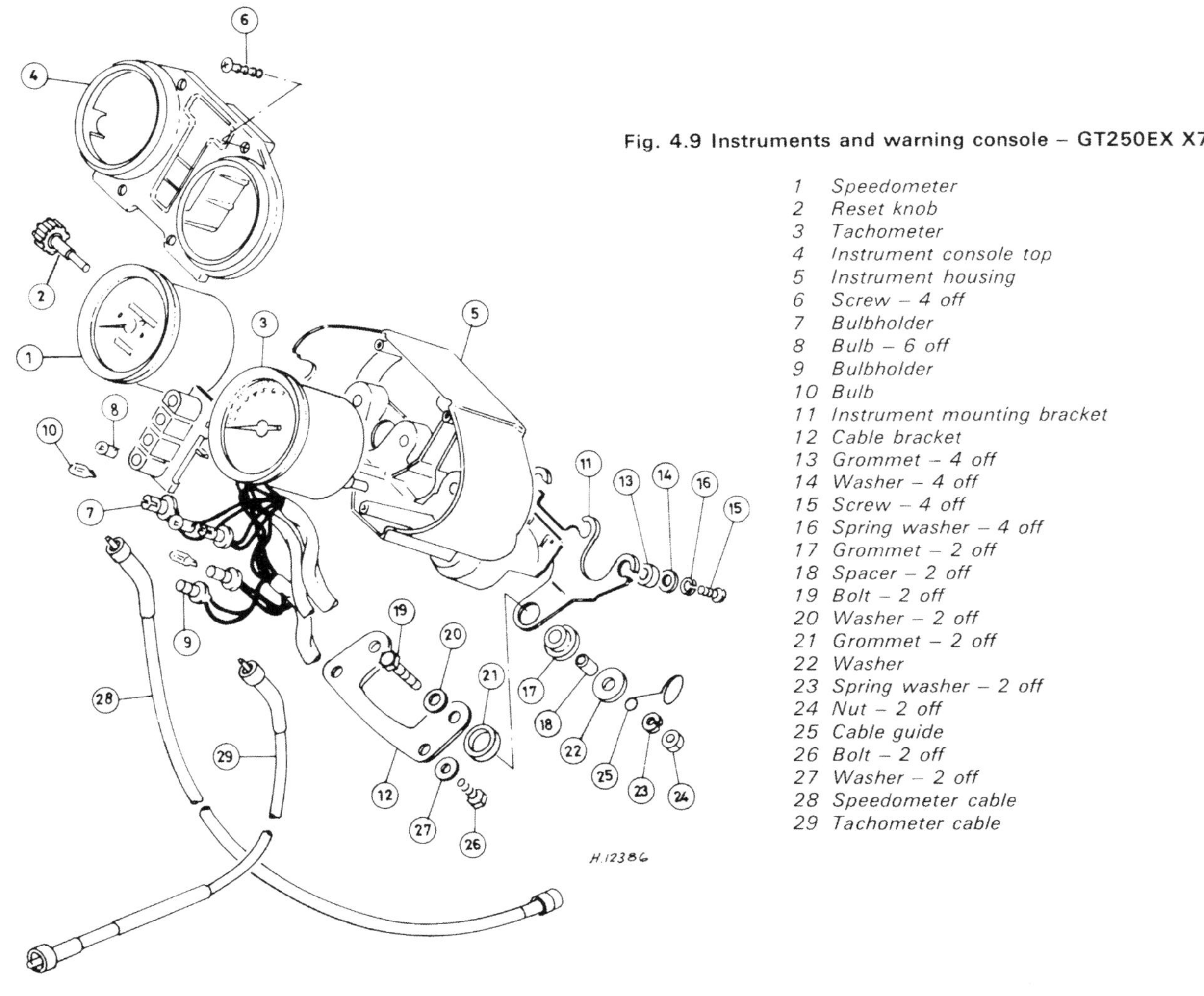

Fig. 4.9 Instruments and warning console – GT250EX X7

1 Speedometer
2 Reset knob
3 Tachometer
4 Instrument console top
5 Instrument housing
6 Screw – 4 off
7 Bulbholder
8 Bulb – 6 off
9 Bulbholder
10 Bulb
11 Instrument mounting bracket
12 Cable bracket
13 Grommet – 4 off
14 Washer – 4 off
15 Screw – 4 off
16 Spring washer – 4 off
17 Grommet – 2 off
18 Spacer – 2 off
19 Bolt – 2 off
20 Washer – 2 off
21 Grommet – 2 off
22 Washer
23 Spring washer – 2 off
24 Nut – 2 off
25 Cable guide
26 Bolt – 2 off
27 Washer – 2 off
28 Speedometer cable
29 Tachometer cable

19 Fault diagnosis: frame and forks

Symptom	Cause	Remedy
Machine vears either to the left or the right with hands off handlebars	Bent frame	Check and renew.
	Twisted forks	Check and renew.
	Wheel out of alignment	Check and re-align.
Machine rolls at low speed	Overtight steering head bearings	Slacken until adjustment is correct.
Machine judders when front brake is applied	Slack steering head bearings	Tighten until adjustment is correct.
	Worn fork sliders	Dismantle forks and renew lower legs and/or stanchions.
Machine pitches on uneven surfaces	Ineffective fork dampers	Check oil content.
	Ineffective rear suspension units	Check whether units still have damping action.
	Suspension too soft	Raise suspension unit adjustment one notch.
Fork action stiff	Fork legs out of alignment (twisted in yokes)	Slacken yoke clamps, and fork top bolts. Pump fork several times then retighten from bottom upwards.
Machine wanders. Steering imprecise. Rear wheel tends to hop	Worn swinging arm pivot	Dismantle and renew bushes (bearings, GT250 X7) and pivot shaft.

Chapter 5 Wheels, brakes and tyres

Contents

Specifications

	GT250 X7	GT200 X5	SB200
Tyres			
Front	3.00S18–4PR	2.75–18–4PR	2.75–18–4PR
Rear	3.50 S18–4PR	3.00–18–4PR	3.00–18–4PR
Pressures			
Standard:			
Front	21 psi (1.5 kg/cm^2)	25 psi (1.75 kg/cm^2)	25 psi (1.75 kg/cm^2)
Rear	25 psi (1.75 kg/cm^2)	28 psi (2.00 kg/cm^2)	28 psi (2.00 kg/cm^2)
With pillion:			
Front	21 psi (1.5 kg/cm^2)	25 psi (1.75 kg/cm^2)	25 psi (1.75 kg/cm^2)
Rear	32 psi (2.25 kg/cm^2)	32 psi (2.25 kg/cm^2)	32 psi 2.25 kg/cm^2)
High speed-solo:			
Front	25 psi (1.75 kg/cm^2)	–	–
Rear	32 psi (2.25 kg/cm^2)	–	–
High speed-pillion:			
Front	25 psi (1.75 kg/cm^2)	–	–
Rear	35 psi (2.50 kg/cm^2)	–	–
Brakes			
Front	Single hydraulic disc.		sls drum
Rear	Single leading shoe (sls) drum brake		

1 General description

All three machines are equipped with 18 in wheels front and rear, the tyre section fitted depending on the model. The X7 and X5 are fitted with cast alloy wheels, the rear wheel incorporating the rear brake drum. The X7 and X5 are each equipped with an hydraulically operated front disc brake, in line with their sporting image. The more conservatively styled SB200 employs a single leading shoe (sls) drum brake front and rear, and is equipped with wire spoked wheels with chromium plated steel rims.

2 Front wheel: examination and renovation (spoked wheel models)

1 Place the machine on the centre stand so that the front wheel is raised clear of the ground. Spin the wheel and check the rim alignment. Small irregularities can be corrected by tightening the spokes in the affected area, although a certain amount of practice is necessary to prevent over-correction. Any flats in the wheel rim should be evident at the same time. These are more difficult to remove and in most cases it will be necessary to have the wheel rebuilt on a new rim. Apart from the effect on stability, a flat will expose the tyre bead and walls to greater risk or damage.

2 Check for loose or broken spokes. Tapping the spokes is the best guide to tension. A loose spoke will produce quite a different sound and should be tightened by turning the nipple in an anti-clockwise direction. Always re-check for run-out by spinning the wheel again. If the spokes have to be tightened an excessive amount, it is advisable to remove the tyre and tube by the procedure detailed in Section 18 of this Chapter; this is so that the protruding ends of the spokes can be ground off, to prevent them from chafing the inner tube and causing punctures.

3 Front wheel: examination and renovation (cast alloy wheel models)

1 Carefully check the complete wheel for cracks and chipping, particularly at the spoke roots and the edge of the rim. As a general rule a damaged wheel must be renewed as cracks will cause stress points which may lead to sudden failure under heavy load. Small nicks may be radiused carefully with a fine file and emery paper (No. 600-No. 1000) to relieve the stress. If there is any doubt as to the condition of a wheel, advice should be sought from a Suzuki repair specialist.

2 Each wheel is covered with a coating of lacquer, to prevent corrosion. If damage occurs to the wheel and the lacquer finish is penetrated, the bared aluminium alloy will soon start to corrode. A whitish grey oxide will form over the damaged area, which in itself is a protective coating. This deposit however, should be removed carefully as soon as possible and a new protective coating of lacquer applied.

3 Check the lateral run out at the rim by spinning the wheel and placing a fixed pointer close to the rim edge. If the maximum run out is greater than 2.0 mm (0.08 in), Suzuki recommend that the wheel be renewed. This is, however, a counsel of perfection; a run out somewhat greater than this can probably be accommodated without noticeable effect on steering. No means is available for straightening a warped wheel without resorting to the expense of having the wheel skimmed on all faces. If warpage was caused by impact during an accident, the safest measure is to renew the wheel complete. Worn wheel bearings may cause rim run out. These should be renewed as described in Section 12 of this Chapter.

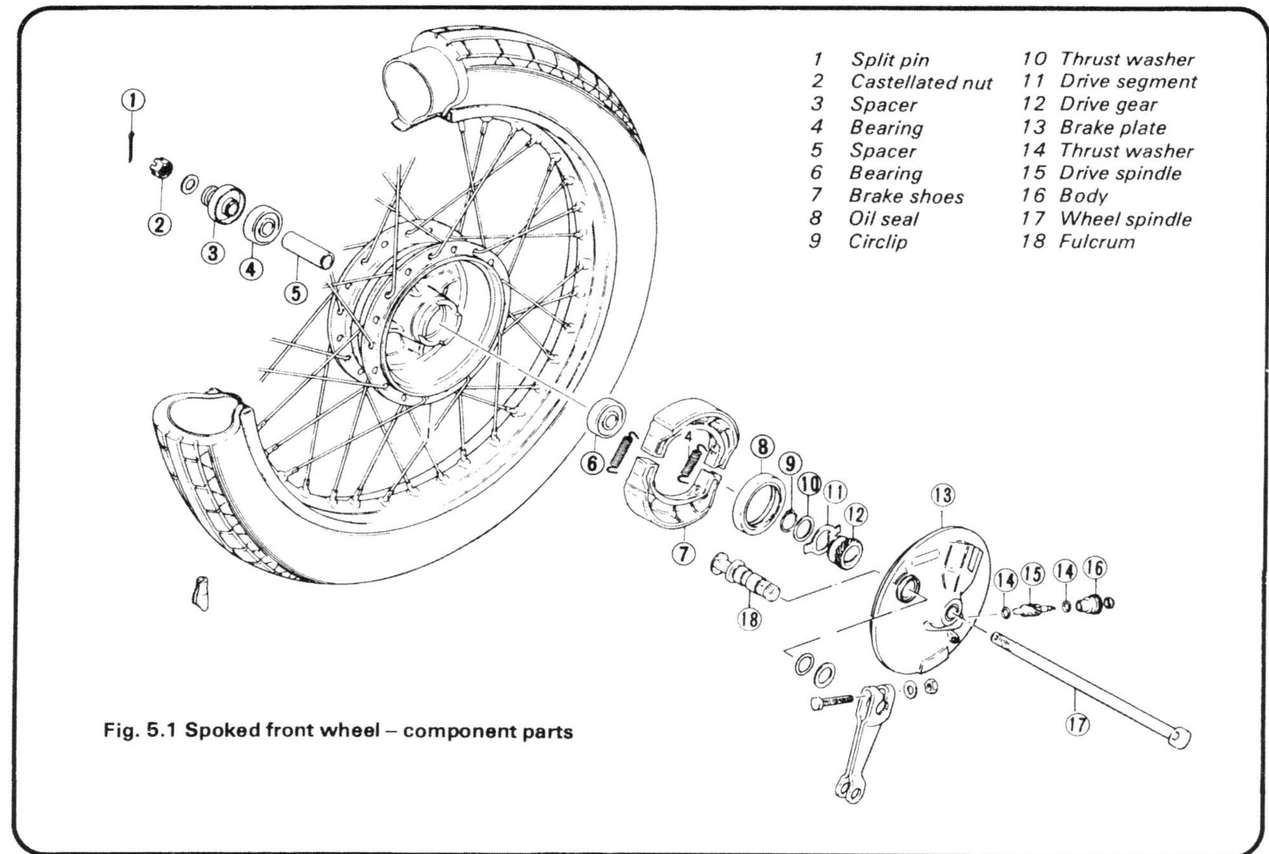

Fig. 5.1 Spoked front wheel – component parts

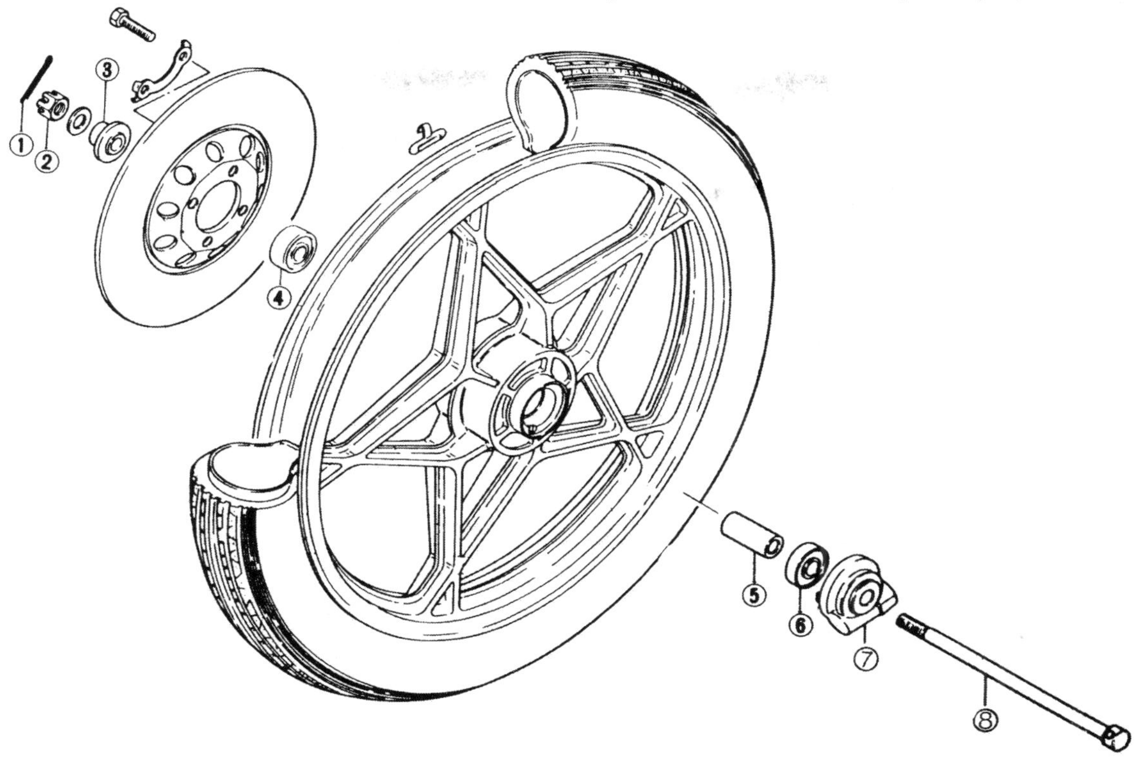

Fig. 5.2 Cast front wheel – component parts

1 Split pin
2 Castellated nut
3 Spacer
4 Bearing
5 Spacer
6 Bearing
7 Speedometer drive gearbox
8 Wheel spindle

4 Front disc brake: removing and replacing the disc and pads

1 The brake disc, attached to the right-hand side of the front wheel hub by four bolts, rarely requires attention. Check for run out, which may have occurred as the result of crash damage, and for wear. Run out should not exceed 0.3 mm (0.01 in) at any point and the disc itself must not be permitted to wear below the limit thickness of 4.5 mm (0.18 in). If these figures are exceeded in either case, the disc must be renewed.
2 The disc is secured directly to the front wheel hub by four bolts, these being locked in pairs by double tab washers. The disc can be removed after the front wheel has been released from the front forks. Place the machine securely on its centre stand so that the front wheel is raised clear of the ground. Withdraw the split pin from the wheel spindle end, and slacken the castellated nut whilst holding the wheel spindle with a tommy bar. Before removing the wheel spindle, release the speedometer drive cable. Support the front wheel with one hand, then twist the wheel spindle out of position. It is advisable to insert a thin wooden wedge between the disc brake pads, to prevent their being expelled if the brake lever is inadvertently squeezed.
3 With the wheel and disc removed, the pads can be examined for wear or damage. Alternatively, the pads and caliper can be dealt with whilst the wheel is installed, by releasing the caliper from its mounting. If the latter course of action is selected, it may be found more convenient to release the caliper from its support bracket by removing the two large Allen bolts on which it slides. This gives better access than removing the caliper and support bracket from the fork leg as an assembly.
4 The caliper is of the floating type, having one fixed, and one moving pad. Slacken the screw which retains the fixed pad, remove its backing plate, and displace the pad inwards. The moving pad should now be dislodged by careful use of a small screwdriver, and withdrawn through the hole previously occupied by the fixed pad. In practice, the moving pad has a tendency to seize in the support bracket which locates it. This is probably due to corrosion from road salt during winter, and electrolytic action between the support bracket and the pad backing metal. This area should be checked carefully and any corrosion carefully removed, as it can seriously impair brake operation.
5 The pads should be examined carefully for wear. Each pad has a red line inscribed around it, denoting the maximum wear limit. This can be observed without dismantling work and makes routine checks simple and quick, but it is desirable to remove the pads from time to time for a more thorough examination. When reassembling the pads, the metal edge of the moving pad should be given a very light coating of brake pad grease. Take care not to allow grease to contaminate the friction material, and avoid using normal grease in this area. Do not omit the anti-squeal shim behind the moving pad (X5 model only).
6 It is important that the caliper assembly is kept scrupulously clean, and any accumulation of road dirt should be removed using a rag moistened with clean brake fluid. Do not use other solvents, as these may attack the brake seals, causing failure at a later stage. After reassembly, check that the brake operates efficiently. If any discrepancy is detected, identify and resolve the problem before using the machine on the road.

4.1 Disc is retained by four bolts and tabwashers

4.2a Remove split pin and wheel spindle nut ...

4.2b ... and remove spindle, not forgetting ...

4.2c ... to release speedometer cable

4.2d Note flanged spacer fitted on disc side of wheel

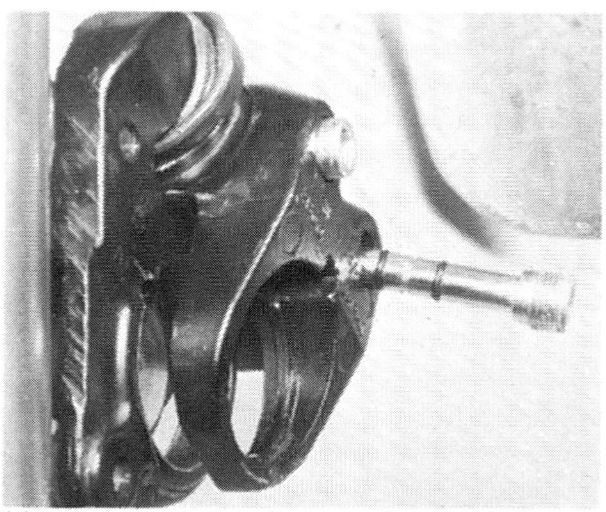
4.3a Release Allen bolts to allow removal of caliper body ...

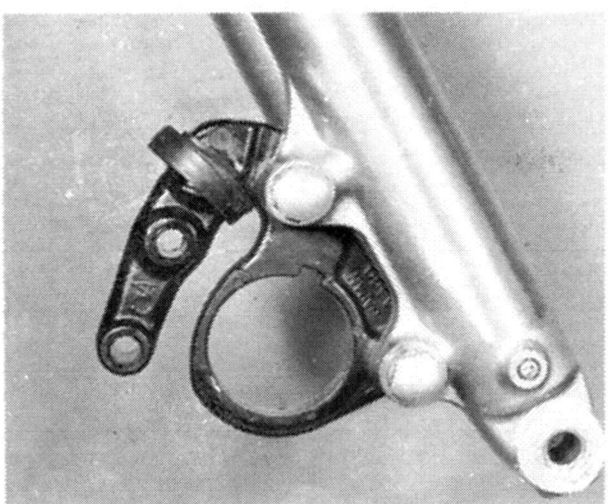
4.3b ... leaving bracket attached to fork leg

4.4a Remove fixed pad ...

4.4b ... then displace moving pad

4.4c Clean bracket surface, and apply a trace of brake grease

4.5a Line on pad denotes maximum wear limit

4.5b Use flat bar to push moving pad into place

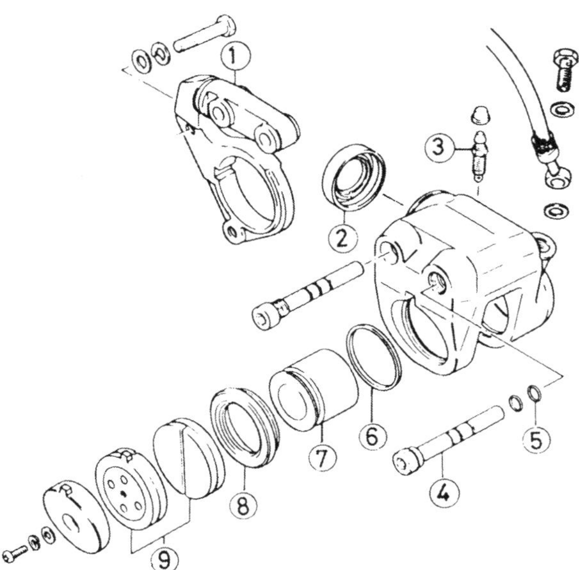

Fig. 5.3 Front brake caliper

1	*Bracket*	6	*Piston seal*
2	*Dust seal*	7	*Piston*
3	*Bleed screw*	8	*Dust seal*
4	*Allen-headed bolts*	9	*Pads*
5	*O-rings*		

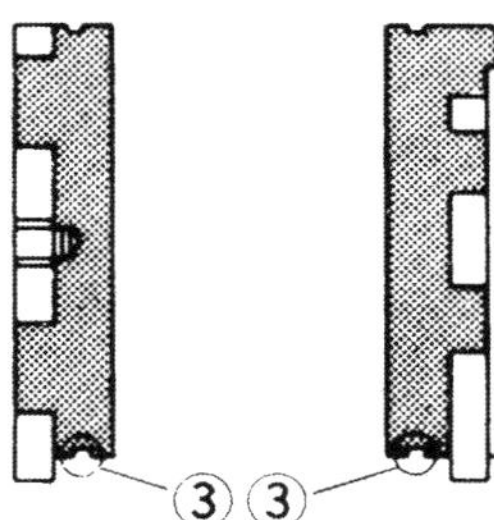

Fig. 5.4 Disc pad wear limit marks
Groove 3 denotes maximum wear limit

5 Front disc brake: removing, renovating and replacing the caliper unit

1 The caliper unit is of the floating type. That is, the caliper body has a controlled sliding motion to allow the fixed pad to be brought into contact with the disc, under pressure, from the opposing, moving, pad. This movement is effected by allowing the body to slide, via two pins, in relation to the mounting bracket. The mounting bracket and body can be separated after unscrewing the two Allen headed bolts on which the caliper slides. If wear develops in this area, it can lead to judder under braking, and the sliding parts should be cleaned and checked carefully.

2 If the caliper unit warrants removal for inspection or renovation, it is first necessary to remove and drain the hydraulic hose. Disconnect the union at the caliper. Have a suitable container in which to catch the fluid. At this stage, it is as well to stop the flow of fluid from the reservoir, by holding the front brake lever in against the handlebar. This is easily done using a stout elastic band, or alternatively, a section cut from an old inner tube.

3 Note: Brake fluid will discolour or remove paint if contact is allowed. Avoid this where possible and remove accidental splashes immediately. Similarly, avoid contact between the fluid and plastic parts such as instrument lenses, as damage will also be done to these. When all the fluid is drained from the hose, clean the connections carefully and secure the hose end and fittings inside a clean polythene bag, to await reassembly. As with all hydraulic systems, it is most important to keep each component scrupulously clean, and to prevent the ingress of any foreign matter. For this reason, it is as well to prepare a clean area in which to work, before further dismantling. As in any form of component dismantling, ensure that the outside of the caliper is thoroughly cleaned down.

4 The caliper unit is attached to the inside of the right-hand fork leg by two bolts, which, when removed, will allow the unit to be lifted away. If the caliper is being removed with the front wheel in position, it should be lifted clear of the disc. Remove the brake pads as described in the preceding Section, exposing the piston. The piston may be driven out of the caliper body by an air jet – a foot pump if necessary. Remove the piston seal and dust seal from the caliper body. Under no circumstances should any attempt be made to lever or prise the piston out of the caliper. If the compressed air method fails, temporarily reconnect the caliper to the flexible hose, and use the handlebar lever to displace the piston hydraulically. Wrap some rag around the caliper to catch the inevitable shower of brake fluid.

5 Clean each part carefully, using only clean hydraulic fluid. On no account use petrol, oil or paraffin as these will cause the seals to degrade and swell. Keep all components dust free.

6 Examine the piston surface for scoring or pitting, any imperfection will necessitate renewal. The seals should be renewed as a matter of course, re-using an old seal is a false economy. Remember that the safety of the machine is very much dependent on seal and piston condition.

7 Reassembly, again ensuring absolute cleanliness, by reversing the dismantling procedure. Use clean hydraulic fluid as lubricant. Replace the caliper unit on the machine and reconnect the hydraulic hose. Remember that the system will need bleeding before use, by following the instructions given in Section 8 of this Chapter.

6 Front disc brake: master cylinder examination and renovation

1 The master cylinder assembly comprises the clear plastic fluid reservoir, the master cylinder body and the brake lever, and is clamped to the handlebar right-hand end. GT250EN and GT200EN models are fitted with round-bodied master cylinders made either by Nissin or by Asco, while GT250EX and GT200EX models are fitted with square-bodied master cylinders made either by Nissin or by Aishin. Differences between the two makes of each type are minimal; refer to the appropriate figure before commencing work. If replacement parts are required, it is essential to specify the exact type and make of master cylinder fitted when ordering.

2 The unit must be drained before any dismantling can be undertaken. Place a container below the caliper unit and run a length of plastic tubing from the caliper bleed screw to the container. Unscrew the bleed screw one full turn and proceed to empty the system by squeezing the front brake lever. When all the fluid has been expelled, tighten the bleed screw and remove the tube.

3 Remove the front brake lever pivot bolt, and lift away the lever. Detach the hose by undoing the banjo mounting at the end of the cylinder. Remove the master cylinder cap – or cover and the diaphragm. Remove the two bolts holding the unit to the handlebars, and lift away. Place the unit on a sheet of paper to provide a clean working area. Pull off the rubber boot fitted to the end of the cylinder, exposing the circlip. Remove the circlip, using circlip pliers to reach it, and withdraw the piston assembly and spring.

4 Examine the piston and seals for scoring or wear and renew if imperfect. Excessive scoring may be due to contaminated fluid, and if this is suspected, it is probably worth checking the condition of the caliper seals and piston.

5 If it is necessary to remove the reservoir body, displace first the retaining screws; there are two of these on all but the Nissin-made square-bodied type, in which the four cover retaining screws pass through the reservoir and into the cylinder body. Remove the reservoir as though unscrewing it, but note that the sealing O-ring must be renewed whenever the joint is disturbed in this way. Lubricate the new O-ring with hydraulic fluid on reassembly.
6 Reassemble carefully, using hydraulic fluid as a lubricant on seals and piston, reversing the dismantling sequence. Make sure the rubber boot is fitted correctly, and that the unit is clamped securely to the handlebars. Reconnect the hydraulic hose, tightening the banjo union bolt to a torque setting of 1.5 – 2.5 kgf m (11 – 18 lbf ft). Refill the reservoir, remembering to top up after the system has been bled by following the procedure given in Section 8 of this Chapter.

7 Front disc brake: hydraulic hose examination

1 An external hose is employed to transmit the hydraulic pressure from the master cylinder to the caliper unit when pressure is applied to the front brake lever.
2 The hose, of the flexible armoured type, must withstand considerable pressure in service, and whilst it is easily ignored, it should be checked carefully as a sudden failure can be potentially fatal. Look not only for signs of chafing against the wheel or fork leg, but also for any stains due to fluid seepage from cracks in the hose or from the connections at either end.

8 Front disc brake: bleeding the hydraulic system

1 Removal of all the air from the hydraulic system is essential to the efficiency of the braking system. Air can enter the system due to leaks or when any part of the system has been dismantled for repair or overhaul. Topping the system up will not suffice, as air pockets will still remain, even small amounts causing dramatic loss of brake pressure.
2 Check the level in the reservoir, and fill almost to the top. Again, beware of spilling the fluid on to painted or plastic surfaces.
3 Place a clean jar below the brake caliper unit and attach a clear plastic tube from the caliper bleed screw to the container. Place some clean hydraulic fluid in the container so that the pipe is always immersed below the surface of the fluid.
4 Unscrew the bleed screw one complete turn and pump the handlebar lever slowly. As the fluid is ejected from the bleed screw the level in the reservoir will fall. Take care that the level does not drop too low whilst the operation continues, otherwise air will re-enter the system, necessitating a fresh start.
5 Continue the pumping action with the lever until no further air bubbles emerge from the end of the plastic pipe. Hold the brake lever against the handlebars and tighten the caliper bleed screw. Remove the plastic tube after the bleed screw is closed.
6 Check the brake action for sponginess, which usually denotes there is still air in the system. If the action is spongy, continue the bleeding operation in the same manner, until all traces of air are removed.
7 When all traces of air have been removed from the system, top up the reservoir and refit the diaphragm and cap. Check the entire system for leaks, and check also that the brake system in general is functioning efficiently before using the machine on the road.
8 Brake fluid drained from the system will almost certainly be contaminated, either by foreign matter or more commonly by the absorption of water from the air. All hydraulic fluids are to some degree hygroscopic, that is, they are capable of drawing water from the atmosphere, and thereby degrading their specifications. In view of this, and the relative cheapness of the fluid, old fluid should always be discarded.

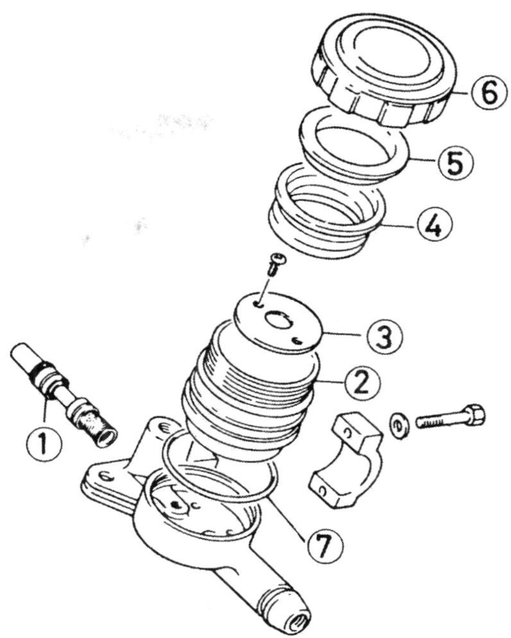

Fig. 5.5 Master cylinder – GT250EN and GT200EN

1 *Piston and cup assembly*
2 *Reservoir*
3 *Plate (Nissin only)*
4 *Diaphragm*
5 *Retainer*
6 *Cap*
7 *O-ring*

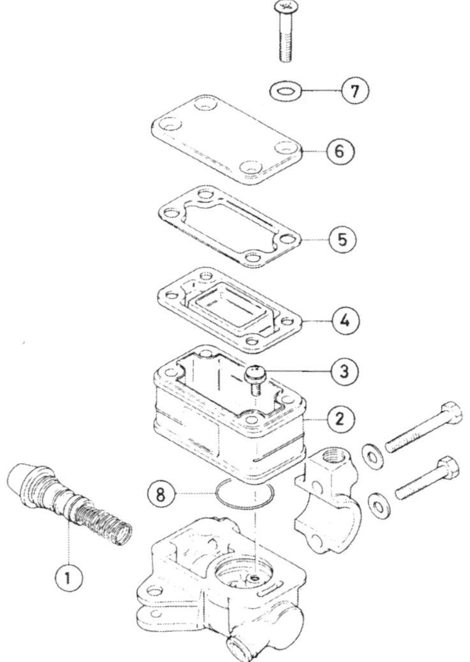

Fig. 5.6 Master cylinder – GT250EX and GT200EX

1 *Piston and cup assembly*
2 *Reservoir*
3 *Screw – 2 off (Aishin only)*
4 *Diaphragm*
5 *Retainer*
6 *Cover*
7 *Washer – 4 off (Nissin only)*
8 *O-ring*

8.3 Clear plastic tube fitted to caliper bleed nipple

8.7 Ensure that fluid level is maintained

9 Front drum brake: examination and renovation – SB200

1 Remove the front wheel from the forks. This is accomplished in the same manner as described for disc brake wheels (see Section 4.2) with the obvious exception that the brake operating cable must be disconnected first. With the wheel removed, the brake plate assembly can be lifted away for examination.
2 Examine the drum surface for signs of scoring or oil contamination. Both of these conditions will impair braking efficiency. Remove all traces of dust, preferably using a brass wire brush, taking care not to inhale any of it, as it is of an asbestos nature, and consequently hazardous. Remove oil or grease deposits, using a petrol soaked rag.
3 If deep scoring is evident, due to the linings having worn through to the shoe at some time, the drum must be skimmed on a lathe or renewed. Whilst there are firms who will undertake to skim a drum whilst fitted to the wheel, it should be borne in mind that excessive skimming will change the radius of the drum in relation to the brake shoes, therefore reducing the friction area until extensive bedding in has taken place. Also full adjustment of the shoes may not be possible. If in doubt about this point, the advice of one of the specialist engineering firms who undertake this work should be sought.
4 If fork oil or grease from the wheel bearings has badly contaminated the linings, they should be renewed. There is no satisfactory way of degreasing the lining material, which in any case is relatively cheap to replace. It is a false economy to try to cut corners with brake components; the whole safety of both machine and rider being dependent on their condition.
5 The linings are bonded to the shoes, and the shoe must be renewed complete with the new linings. This is accomplished by folding the shoes together until the spring tension is relaxed, and then lifting the shoes and springs off the brake plate. Fitting new shoes is a direct reversal of the above procedure.
6 Before refitting existing shoes, roughen the lining surface sufficiently to break the glaze which will have formed in use.
7 Whilst the shoes are removed, release the brake operating arm and displace the fulcrum. Clean the fulcrum and its bore in the brake plate, and grease both parts to ensure smooth operation.

10 Rear wheel: removal and replacement

1 The procedure for removing and replacing the rear wheel is similar, irrespective of whether cast or spoked wheels are employed. Place the machine on its centre stand. Unscrew the brake operating rod adjuster nut, and depress the pedal to disengage the rod from the operating arm. Reassemble the adjuster components on the operating rod to avoid their subsequent loss. Disconnect the brake torque arm from the brake plate, by removing the spring clip and retaining nut.
2 Withdraw the wheel spindle split pin, and slacken the castellated nut. Support the rear wheel assembly and withdraw the wheel spindle. As the wheel is removed, disengage the sprocket and sprocket hub from the cush drive, leaving the assembly engaged with the drive chain.
3 Reassembly is a direct reversal of the removal sequence. Ensure that the brake and chain tension adjustments are checked before using the machine on the road.

11 Rear drum brake: examination and renovation

1 The procedure for overhauling the rear brake is similar to that described for the front brake of SB200 models in Section 9.

12 Front wheel bearings: examination and renovation

1 The front wheel bearings are an interference fit in the wheel hub, and can be removed by passing a long drift through the centre of one bearing and driving the remaining bearing out from the opposite side. On disc brake wheels, it is advisable to support the wheel on wooden blocks to avoid damage to the disc, or to remove the disc from the wheel.
2 With the wheel suitably supported, pass the drift into position, displacing the spacer between the bearings so that the drift can bear on the inner race of the right-hand bearing. Drive the bearing out of the hub, and remove the spacer.
3 Invert the wheel and drive out the left-hand bearing by inserting a drift of the appropriate size, through the hub. During the removal of either bearing it may be necessary to support the wheel across an open-ended box so that there is sufficient clearance for the bearing to be displaced completely from the hub.
4 Remove all the old grease from the hub and bearings, giving the latter a final wash in petrol. Check the bearings for signs of play or roughness when they are turned. If there is any doubt about the condition of a bearing, it should be renewed.
5 Before replacing the bearings, first pack the hub with new grease. Then drive the bearings back into position, not forgetting the distance piece that separates them. Take great care to ensure that the bearings enter the housings perfectly squarely otherwise the housing surface may be broached.

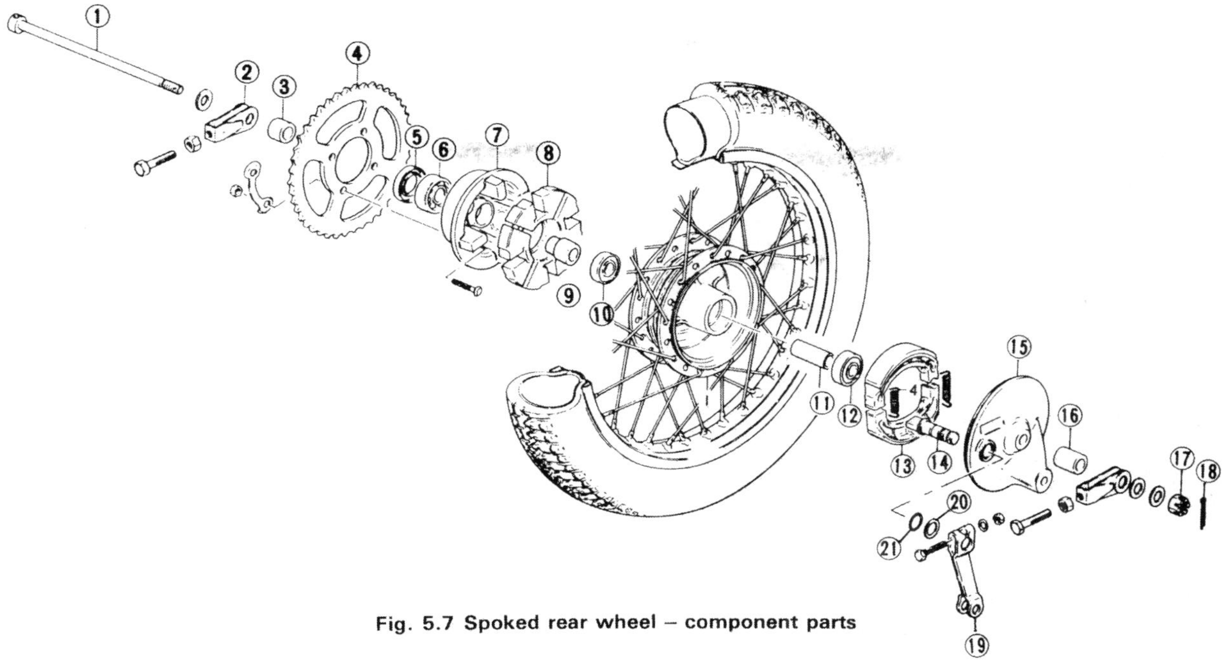

Fig. 5.7 Spoked rear wheel – component parts

1 Wheel spindle
2 Chain adjuster
3 Spacer
4 Sprocket
5 Oil seal
6 Bearing
7 Sprocket/cush drive hub
8 Cush drive rubbers
9 Inner sleeve
10 Bearing
11 Spacer
12 Bearing
13 Brake shoes
14 Fulcrum
15 Brake plate
16 Spacer
17 Castellated nut
18 Split pin
19 Brake arm
20 Washer
21 O-ring

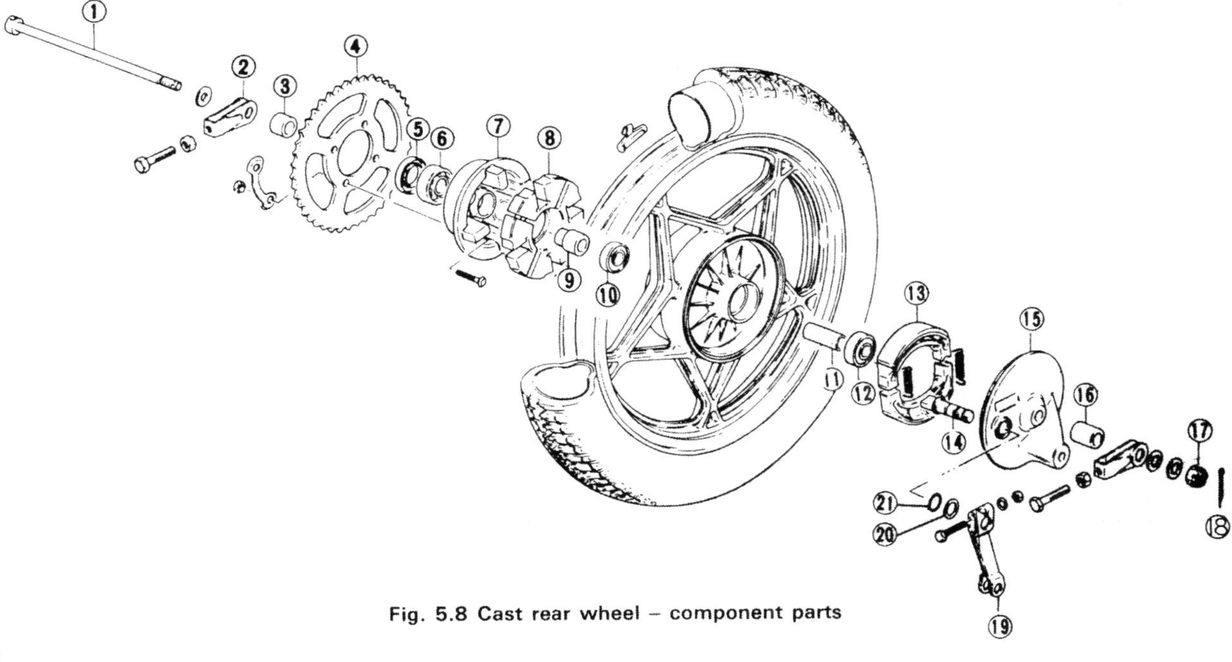

Fig. 5.8 Cast rear wheel – component parts

1 Wheel spindle
2 Chain adjuster
3 Spacer
4 Sprocket
5 Oil seal
6 Bearing
7 Sprocket/cush drive hub
8 Cush drive rubbers
9 Inner sleeve
10 Bearing
11 Spacer
12 Bearing
13 Brake shoes
14 Fulcrum
15 Brake plate
16 Spacer
17 Castellated nut
18 Split pin
19 Brake arm
20 Washer
21 O-ring

10.1a Slacken and remove rear brake adjuster

10.1b Remove R pin and torque arm securing nut

10.2a Withdraw split pin and remove wheel spindle nut

10.2b Note spacer on sprocket side of rear wheel

10.2c Withdraw spindle and displace spacer to release wheel

11.1a Rear brake plate assembly removed for inspection

Tyre changing sequence - tubed tyres

Deflate tyre. After pushing tyre beads away from rim flanges push tyre bead into well of rim at point opposite valve. Insert tyre lever adjacent to valve and work bead over edge of rim.

Use two levers to work bead over edge of rim. Note use of rim protectors

Remove inner tube from tyre

When first bead is clear, remove tyre as shown

When fitting, partially inflate inner tube and insert in tyre

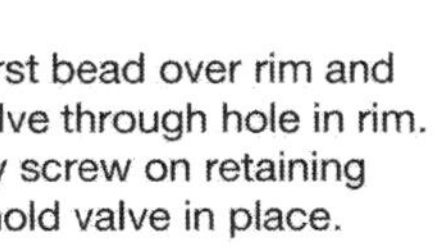

Work first bead over rim and feed valve through hole in rim. Partially screw on retaining nut to hold valve in place.

Check that inner tube is positioned correctly and work second bead over rim using tyre levers. Start at a point opposite valve.

Work final area of bead over rim whilst pushing valve inwards to ensure that inner tube is not trapped

11.1b Shoes are removed by folding them inwards

11.1c Check drum surface for corrosion or scoring

12.1 Front wheel bearings are driven into hub

13 Rear wheel bearings: examination and renovation

1 The rear wheel bearings can be dealt with in the same manner as that described for the front wheel bearings in Section 12 of this Chapter. Note that there is an additional outrigger bearing and separate oil seal contained in the cush drive hub, and this should not be omitted when checking the rear bearings (see Section 16).

14 Speedometer drive gearbox: examination and lubrication

1 The speedometer drive is taken from a gearbox mounted on the front wheel of all models. On drum brake models, the drive gear is supported in the brake plate casting, as is the driven shaft. The drive gear can be removed from the inside of the brake plate, together with the driving dog, after removing the circlip which retains it. The driven shaft can be removed from the outside of the brake plate, by unscrewing its retaining gland nut. Unless excessively worn due to lack of lubricant, which will necessitate renewal, little can be done by way of maintenance other than thorough cleaning and relubrication of the components and their housing.

2 Maintenance of the disc brake type of gearbox should be approached in the same manner, in this case, the unit being a self-contained assembly which is a push fit in the front hub. The unit should be greased as a matter of course, each time the front wheel is removed.

15 Cush drive assembly: examination and renovation

1 The cush drive assembly is contained within the left-hand side of the rear wheel hub. It comprises a set of synthetic rubber buffers, housed within a series of vanes cast in the hub shell. A plate attached to the centre of the rear wheel sprocket has four cast-in dogs which engage with slots in these rubbers, when the wheel is replaced in the frame. The drive to the rear wheel is transmitted via these rubbers, which cushion any surges of roughness in the drive which would otherwise convey the impression of harshness.
2 Examine the rubbers periodically for signs of damage or general deterioration. Renew and fit the rubbers as a set if there is any doubt about their condition; there is no difficulty in removing or replacing them as they are not under compression when the drive plate is attached.

13.1a Remove bearings using long drift passed through hub

13.1b Bearings on both sides of hub ...

13.1c ... have integral oil seals

13.1d Cush drive carries outrigger bearing ...

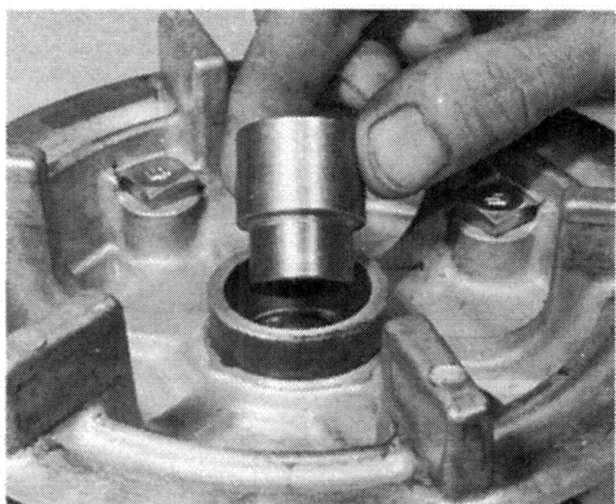

13.1e ... which locates round short outer spindle

14.1a Tangs on speedometer drive locate in slots in hub

14.1b Spindle is retained by sleeve nut

15.1 Remove sprocket/cush drive hub ...

15.2 ... and check damping rubbers for wear

16 Rear wheel sprocket: removal, examination and replacement

1 The rear wheel sprocket assembly can be removed as a separate unit after the rear wheel has been separated and detached from the frame as described in Section 10 of this Chapter. Alternatively, it can be removed still attached to the rear wheel if the final drive chain is detached.

2 Check the condition of the sprocket teeth. If they are hooked, chipped or badly worn, the sprocket must be renewed. It is secured to the cush drive plate by four bolts and lock washers.

3 It is considered bad practice to renew one sprocket on its own. The final drive sprockets should always be renewed as a pair and a new chain fitted, otherwise rapid wear will necessitate even earlier renewal on the next occasion.

4 An additional bearing is located within the cush drive plate, which supports the sprocket shaft into which the rear wheel spindle fits. In common with the wheel bearings, this bearing is a journal ball and when wear occurs, the sprocket will give the appearance of being loose on its mounting bolts. The bearing is a tight push fit on the sprocket shaft and is preceded by an oil seal that excludes road grit and water.

5 Remove the oil seal and bearing and wash out the latter to remove all traces of the old grease. If the bearing has any play or runs roughly, it must be renewed.

6 If the bearing has not been renewed it should be repacked with grease and refitted in its housing, followed by the oil seal. Replace the rear wheel assembly by reversing whichever method was adopted for its removal.

17 Final drive chain: examination and lubrication

1 The final drive chain is fully exposed, with only a light chainguard over the top run. Periodically the tension will need to be adjusted, to compensate for wear. This is accomplished by placing the machine on the centre stand and slackening the two wheel nuts on the left-hand side of the rear wheel so that the wheel can be drawn backward by means of the drawbolt adjusters in the fork ends. The rear brake torque arm bolt must also be slackened during this operation.

2 The chain is in correct tension if there is approximately 20 mm (¾ in) slack in the middle of the lower run. Always check when the chain is at its tightest point as a chain rarely wears evenly during service.

3 Always adjust the drawbolts an equal amount in order to preserve wheel alignment. The fork ends are clearly marked with a series of horizontal lines above the adjusters, to provide a simple, visual check. If desired, wheel alignment can be checked by running a plank of wood parallel to the machine, so that it touches the side of the rear tyre. If wheel alignment is correct, the plank will be equidistant from each side of the front wheel tyre, when tested on both sides of the rear wheel. It will not touch the front wheel tyre because this tyre is of smaller cross section. See accompanying diagram.

4 Do not run the chain overtight to compensate for uneven wear. A tight chain will place undue stress on the gearbox and rear wheel bearings, leading to their early failure. It will also absorb a surprising amount of power.

5 After a period of running, the chain will require lubrication. Lack of oil will greatly accelerate the rate of wear of both the chain and the sprockets and will lead to harsh transmission. The application of engine oil will act as a temporary expedient, but it is preferable to remove the chain and clean it in a paraffin bath before it is immersed in a molten lubricant such as 'Linklife' or 'Chainguard'. These lubricants achieve better penetration of the chain links and rollers and are less likely to be thrown off when the chain is in motion.

6 The GT250 X7 model is fitted with an 'endless' chain which has no joining link. In consequence of this, chain removal requires that the swinging arm is detached. In view of this cleaning and lubrication of the chain must be carried out whilst it is on the machine, and it is suggested that an aerosol chain lubricant is used in place of the molten type of lubricant.

7 To check whether the chain is due for replacement, lay it lengthwise in a straight line and compress it endwise so that all the play is taken up. Anchor one end and measure the length. Now pull the chain with one end anchored firmly, so that the chain is fully extended by the amount of play in the opposite direction. If there is a difference of more than ¼ inch per foot in the two measurements, the chain should be replaced in conjunction with the sprockets. Note that this check should be made after the chain has been washed out, but before any lubricant is applied, otherwise the lubricant may take up some of the play.

8 When replacing the chain, make sure that the spring link is seated correctly, with the closed end facing the direction of travel.

9 Replacement chains are now available in standard metric sizes from Renold Limited, the British chain manufacturer. When ordering a new chain, always quote the size, the number or chain links and the type of machine to which the chain is to be fitted.

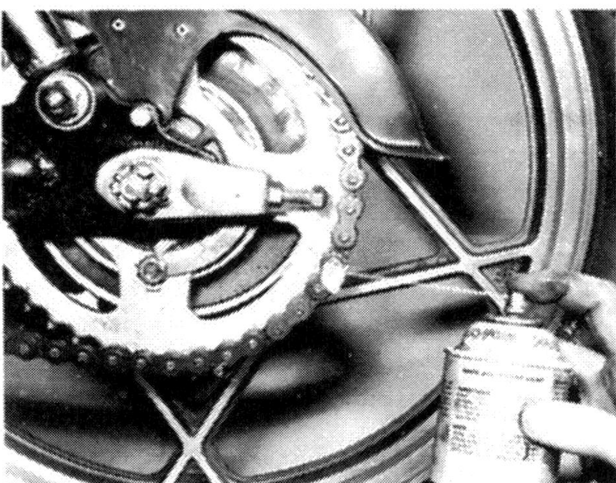

17.6 Endless type chains can only be lubricated by aerosol

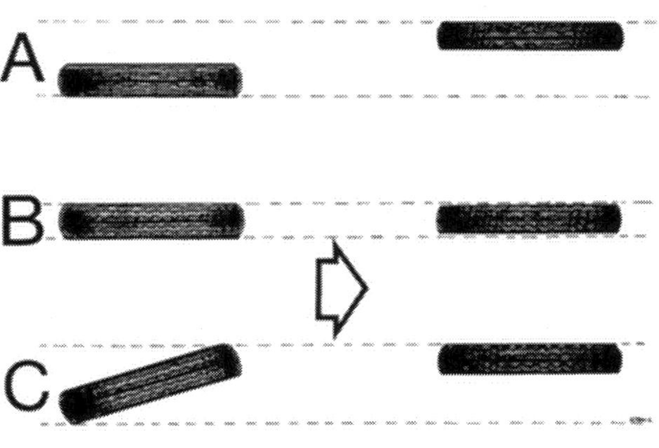

Fig. 5.9 Checking wheel alignment

A and C Incorrect *B correct*

18 Tyres: removal and replacement

1 At some time or other the need will arise to remove and replace the tyres, either as the result of a puncture or because a renewal is required to offset wear. To the inexperienced, tyre changing represents a formidable task yet if a few simple rules are observed and the technique learned, the whole operation is surprisingly simple.

2 To remove the tyre from the wheel, first detach the wheel from the machine by following the procedure in Section 4 or Section 10 of this Chapter, depending on whether the front or the rear wheel is involved. Deflate the tyre by removing the valve insert and when it is fully deflated, push the bead of the tyre away from the wheel rim on both sides so that the bead enters the centre well of the rim. Remove the locking cap and push the valve into the tyre itself.

3 Insert a tyre lever close to the valve and lever the edge of the tyre over the outside of the wheel rim. Very little force should be necessary; if resistance is encountered it is probably due to the fact that the tyre beads have not entered the well of the wheel rim all the way round the tyre.

4 Once the tyre has been edged over the wheel rim, it is easy to work around the wheel rim so that the tyre is completely free on one side. At this stage, the inner tube can be removed.

5 Working from the other side of the wheel, ease the other edge of the tyre over the outside of the wheel rim farthest away. Continue to work around the rim until the tyre is free from the rim.

6 If a puncture has necessitated the removal of the tyre, reinflate the inner tube and immerse it in a bowl of water to trace the source of the leak. Mark its position and deflate the tube. Dry the tube and clean the area around the puncture with a petrol-soaked rag. When the surface has dried, apply rubber solution and allow this to dry before removing the backing from the patch and applying the patch to the surface.

7 It is best to use a patch of the self-vulcanising type, which will form a very permanent repair. Note that it may be necessary to remove a protective covering from the top surface of the patch after it has sealed in position. Inner tubes made from synthetic rubber may require a special type of patch and adhesive, if a satisfactory bond is to be achieved.

8 Before replacing the tyre, check the inside to make sure the agent that caused the puncture is not trapped. Check also the outside of the tyre, particularly the tread area, to make sure nothing is trapped that may cause a further puncture.

9 If the inner tube has been patched on a number of past occasions, or if there is a tear or large hole, it is preferable to discard it and fit a new one. Sudden deflation may cause an accident, particularly if it occurs with the front wheel.

10 To replace the tyre, inflate the inner tube sufficiently for it to assume a circular shape but only just. Then push it into the tyre so that it is enclosed completely. Lay the tyre on the wheel at an angle and insert the valve through the rim tape and the hole in the wheel rim. Attach the locking cap on the first few threads, sufficient to hold the valve captive in its correct location.

11 Starting at the point furthest from the valve, push the tyre bead over the edge of the wheel rim until it is located in the central well. Continue to work around the tyre in this fashion until the whole of one side of the tyre is on the rim. It may be necessary to use a tyre lever during the final stages.

12 Make sure there is no pull on the tyre valve and again commencing with the area furthest from the valve, ease the other bead of the tyre over the edge of the rim. Finish with the area close to the valve, pushing the valve up into the tyre until the locking cap touches the rim. This will ensure the inner tube is not trapped when the last section of the bead is edged over the rim with a tyre lever.

13 Check that the inner tube is not trapped at any point. Reinflate the inner tube, and check that the tyre is seating correctly around the wheel rim. There should be a thin rib moulded around the wall of the tyre on both sides, which should be equidistant from the wheel rim at all points. If the tyre is unevenly located on the rim, try bouncing the wheel when the tyre is at the recommended pressure. It is probable that one of the beads has not pulled clear of the centre well.

14 Always run the tyres at the recommended pressures and never under or over-inflate. The correct pressures for solo use are given in the Specifications section of this Chapter.

15 Tyre replacement is aided by dusting the side walls, particularly in the vicinity of the beads, with a liberal coating of french chalk. Washing-up liquid can also be used to good effect, but this has the disadvantage of causing the inner surfaces of the wheel rim to rust.

16 Never replace the inner tube and tyre without the rim tape in position (wire wheels only). If this precaution is overlooked there is a good chance of the ends of the spoke nipples chafing the inner tube and causing a crop of punctures.

17 Never fit a tyre that has a damaged tread or side walls. Apart from the legal aspects, there is a very great risk of a blowout which can have serious consequences on any two-wheel vehicle.

18 Tyre valves rarely give trouble, but it is always advisable to check whether the valve itself is leaking before removing the tyre. Do not forget to fit the dust cap, which forms an effective second seal.

19 Valve cores and caps

1 Valve cores seldom give trouble, but do not last indefinitely. Dirt under the seating will cause a puzzling 'slow-puncture'. Check that they are not leaking by applying spittle to the end of the valve and watching for air bubbles.
2 A valve cap is a safety device, and should always be fitted. Apart from keeping dirt out of the valve, it provides a second seal in case of valve failure, and may prevent an accident resulting from sudden deflation.

20 Front wheel balancing

1 The front wheel should be statically balanced, complete with tyre. An out of balance wheel can produce dangerous wobbling at high speed.
2 Some tyres have a balance mark on the sidewall. This must be positioned adjacent to the valve. Even so, the wheel still requires balancing.
3 With the front wheel clear of the ground, spin the wheel several times. Each time, it will probably come to rest in the same position. Balance weights should be attached diametrically opposite the heavy spot, until the wheel will not come to rest in any set position, when spun.
4 Balance weights, which clip round the spokes, are available in 20 or 30 gramme weight. If they are not available, wire solder wrapped around the spokes and secured with insulating tape will make a substitute.
5 It is possible to have a wheel dynamically balanced at some dealers. This requires its removal.
6 There is no need to balance the rear wheel under normal road conditions, although any tyre balance mark should be aligned with the valve.
7 Machines fitted with cast aluminium wheels require special balancing weights which are designed to clip onto the centre rim flange, much in the way that weights are affixed to car wheels.

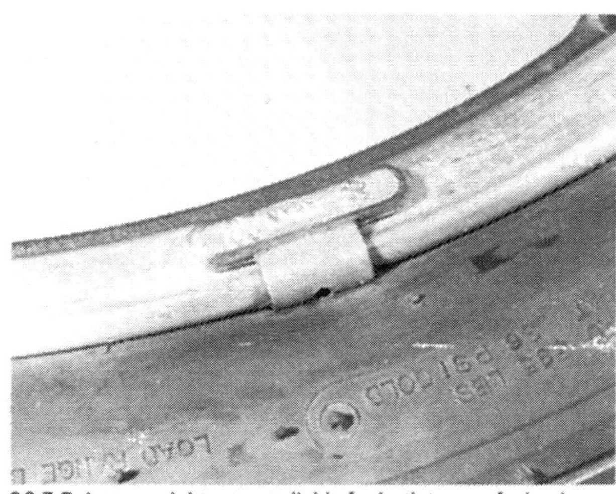
20.7 Balance weights are available for both types of wheel

21 Fault diagnosis: wheels, brakes and tyres

Symptom	Cause	Remedy
Handlebars oscillate at low speeds	Buckled front wheel	Remove wheel for specialist attention. Renew wheel (cast alloy type).
	Incorrectly fitted front tyre	Check whether line around bead is equidistant from rim.
Fork 'hammers' at high speeds	Front wheel out of balance	Add weights until wheel will stop in any position.
Brakes feel spongy	Air in hydraulic line	Bleed brakes.
	Fluid leak in system	Replace faulty part.
Tyres wear more rapidly in middle of tread	Over-inflation	Check pressures and run at recommended settings.
Tyres wear rapidly at outer edges of tread	Under-inflation	Check pressures and run at recommended settings.

Chapter 6 Electrical system

Contents

Specifications

	GT250 X7	GT200 X5	SB200
Battery			
Make	Yuasa	Yuasa	Yuasa
Type	12N5–3B	YB7–A Yumicron	12N5–3B(S)
Voltage	12 volts	12 volts	12 volts
Capacity	5 amp hour	8 amp hour	5 amp hour
Alternator			
Type	Flywheel generator		
Charging system			
Output (daytime)	15–16V	15–16V	15–16V
Output (with lights)	14–15V	14–15V	15–16V
Regulator/rectifier			
Type	Combined unit, with full-wave rectification and Zener diode		
Bulbs			
Headlamp	35/35W	35/35W	35/35W
Parking lamp	3.4W	3.4W	3.4W
Stop/tail lamp	21/5W	21/5W	21/5W
Speedometer lamp	3.4W	3.4W	3.4W
Tachometer lamp	3.4W	3.4W	–
Direction indicators	21W	21W	21W
Neutral lamp	3.4W	3.4W	3.4W
Direction indicator warning lamp	3.4W	3.4W	3.4W
Main beam warning lamp	3.4W	3.4W	1.7W

1 General description

1 The electrical system on the Suzuki X7, X5 and SB200 machines is powered by a flywheel generator mounted on the left-hand end of the crankshaft. Output from the ac generator passes to a combined rectifier/regulator unit, which converts it to direct current (dc).

2 The GT200 X5 model is equipped with an electric starter system powered by a motor contained in a recess in the underside of the crankcase. This equipment is fitted to the SB200 models marketed in a number of European countries, but, at the time of writing, not to the UK models.

2 Charging system: checking the output

1 If the battery appears to be consistently over-charged, indicated by a frequent need to top up the battery, or under-charged, the charging system output should be checked. Obtain a volt meter, or multimeter set on dc volts, capable of measuring 0 – 20 volts. Connect the red probe lead to the positive (+) battery terminal and the black probe lead to the negative (–) battery terminal.

2 Start the engine, and run it up to 5000 rpm, noting the reading obtained with the lights off. Repeat the test with the

lights on. With lights off, a reading of 15 – 16 volts should be shown, and a reading of 14 – 15 volts should be shown with the lights on. If a reading below these figures is obtained, attention should be directed to the switches, connections, charging coils or the regulator/rectifier unit. If a higher value is shown, the regulator/rectifier unit is probably at fault.

3 Regulator/rectifier unit: testing

1 Separate the connector to the unit, and check the resistances between the various terminals, using an ohmmeter or multimeter, following the table given below. If incorrect readings are shown, take the unit to a Suzuki Service Agent for verification of the fault and renewal. If a multimeter is used it should be set on the resistance (ohms) setting.

		Positive probe connections			
		R/G	*Y/G*	*B/W*	*R*
	R/G	–	*INF*	*INF*	*CON*
Negative probe connections	*Y/G*	*INF*	–	*INF*	*CON*
	B/W	*CON*	*CON*	–	*CON*
	R	*INF*	*INF*	*INF*	–

Key:

INF	*infinite resistance, CON: continuity*
R/G	*Red with green tracer*
Y/G	*Yellow with green tracer*
B/W	*Black with white tracer*
R	*Red*

4 Charging coils: testing

1 The charging coils can be checked by measuring their resistances with an ohmmeter or a multimeter set on resistance. Trace the output leads and separate them at the connector, then check the resistance between the various leads, following the tables given below.

X7 models

Connections	*Reading*
Green/White to Yellow/Green	*1 ohm*
Red/green to Yellow/green	*1 ohm*

X5 and SB200 models

Connections	*Reading*
Green/white to Red/green	*2 ohms*
Red/green to Yellow/green	*1 ohm*
Green/white to Yellow/green	*1 ohm*
Green/white to earth	*infinity*
Yellow/green to earth	*infinity*
Red/green to earth	*infinity*

5 Charging system: component renewal

1 The test procedures outlined in the preceding sections should enable a suspect component to be identified in the event of a malfunction in the system. In each case, do not condemn the component concerned until it has been checked by a Suzuki dealer, who will be in the best position to judge whether a replacement is required.
2 Other problems in the electrical system can give rise to charging problems. For example, a damaged or worn out battery will be unable to hold a full charge, and dirty switch contacts will not pass the required amount of current without causing an appreciable amount of resistance. These areas should be checked when an electrical fault is encountered, as described later in this Chapter.

6 Battery: examination and maintenance

1 A conventional Yuasa lead-acid battery is fitted to the Suzuki X7, X5 and SB200 models. On electric start models, a 12 volt 8 amp hour battery is employed, non-electric start machines having a 12 volt 5 amp hour unit.
2 The transparent case of the battery allows the upper and lower levels of the electrolyte to be observed without need to remove the battery. Maintenance is normally limited to keeping the electrolyte level between the prescribed upper and lower limits and making sure the vent tube is not blocked. The lead plates and their separators can be seen through the transparent case, a further guide to the condition of the battery.
3 Unless acid is spilt, as may occur if the machine falls over, the electrolyte should always be topped up with distilled water, to restore the correct level. If acid is spilt on any part of the machine, it should be neutralised with an alkali such as washing soda and washed away with plenty of water, otherwise serious corrosion will occur. Top up with sulphuric acid of the correct specific gravity (1·260 – 1·280) only when spillage has occurred. Check that the vent pipe is well clear of the frame tubes or any of the other cycle parts.
4 It is seldom practicable to repair a cracked case because the acid in the joint prevents the formation of an effective seal. It is always best to replace a cracked battery, especially in view of the corrosion that will be caused by acid leakage.
5 If the machine is laid up for a period, it is advisable to remove the battery and give it a 'refresher' charge every six weeks or so from a battery charger. If the battery is permitted to discharge completely, the plates will sulphate and render the battery useless.

7 Battery: charging procedure

1 The battery charging rate should always be one tenth of the battery's rated capacity. The five amp hour battery will therefore require charging at 0·5 amps, and the eight amp hour unit at 0·8 amps. A higher charging rate should be avoided as this will shorten the working life of the battery, and if excessive, may overheat the battery and buckle its plates.
2 Make sure the charger connections to the battery are correct; red to positive and black to negative. It is preferable to remove the battery from the machine during the charging operation and to remove the vent plug from each cell.

8 Fuse: location and replacement

1 A fuse is incorporated in the electrical system to give protection from a sudden overload, as may occur during a short circuit. It is found within a fuse holder that forms part of the wiring snap connections, close to the battery. A transparent plastic bag attached to the wiring carries a spare fuse, for use in an emergency. The fuse is rated at 15 amps.
2 If a fuse blows, it should be replaced, after checking to ensure that no obvious short circuit has occurred. If the second fuse blows shortly afterwards, the electrical circuit must be checked thoroughly, to trace the fault.
3 When a fuse blows whilst the machine is running and no spare is available, a 'get you home' remedy is to remove the blown fuse and wrap it in silver paper before replacing it in the fuseholder. The silver paper will restore electrical continuity by bridging the broken fuse wire. This expedient should never be used if there is evidence of a short circuit or other major electrical fault, otherwise more serious damage will be caused. Replace the blown fuse at the earliest possible opportunity, to restore full circuit protection.

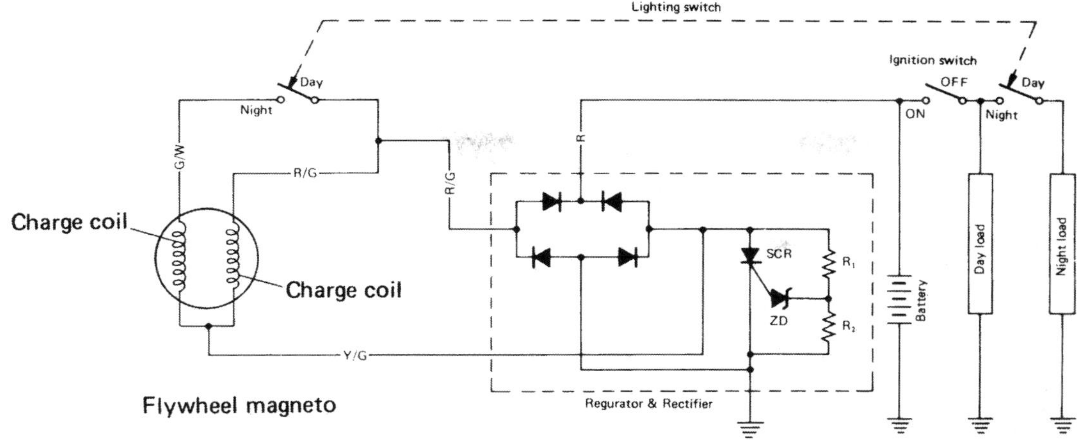

Fig. 6.1 Charging system circuit diagram – GT250 X7 (other models similar)

G/W : Green with white tracer
R/G : Red with green tracer
R : Red
Y/G : Yellow with green tracer
SCR : Silicon Controlled Rectifier
ZD : Zenner diode

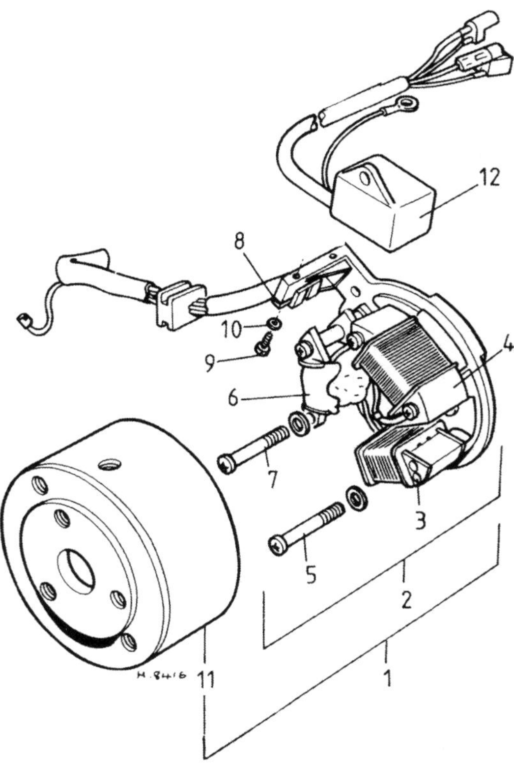

Fig. 6.2 Flywheel generator and CDI unit

1	*Magneto assembly*	*7*	*Screw – 2 off*
2	*Stator assembly*	*8*	*Pulser coil*
3	*Lighting coil A*	*9*	*Screw – 2 off*
4	*Lighting coil B*	*10*	*Spring washer – 2 off*
5	*Screw – 4 off*	*11*	*Flywheel*
6	*Primary coil*	*12*	*C.D.I. unit assembly*

9 Starter motor: removal, examination and replacement

1 An electric starter motor, operated from a small push-button on the right-hand side of the handlebars, provides an alternative and more convenient method of starting the engine, without having to use the kickstart. It is fitted to all GT 200 X5 models, and also the SB 200 models supplied to various European markets, other than the UK. The motor is housed in a recess in the underside of the crankcase. Current is supplied from the battery via a heavy duty solenoid switch and a cable capable of carrying the very high current demanded by the starter motor on the initial start-up.

2 The starter motor drives a free running clutch via an idler pinion. The clutch ensures the starter motor drive is disconnected from the crankshaft immediately the engine starts. It operates on the centrifugal principle; spring loaded rollers take up the drive until the engine starts and the drive is automatically disconnected.

3 The starter motor may be removed while the engine is in the frame. It will be first be necessary to remove the left-hand outer casing and remove the alternator rotor and the idler pinion assembly. Full details of starter motor removal will be found in Section 10 of Chapter 1.

4 The parts of the starter motor most likely to require attention are the brushes. The end cover is retained by the two long screws which pass through the lugs cast on both end pieces. If the screws are withdrawn, the end cover can be lifted away and the brush gear exposed.

5 Lift up the spring clips which bear on the end of each brush and remove the brushes from their holders. Each brush should have a length of 12·0 mm (0·47 in). The minimum allowable brush length is 8·5 mm (0·33 in). If the brush is shorter it must be renewed.

6 Before the brushes are replaced, make sure that the commutator is clean. The commutator is the ring of copper segments on which the brushes bear. Clean the commutator with a strip of glass paper. Never use emery cloth or 'wet-and-dry' as the small abrasive fragments may embed themselves in the soft copper of the commutator and cause excessive wear of the brushes. Finish off the commutator with metal polish to give a smooth surface and finally wipe the segments over with a methylated spirit soaked rag to ensure a grease free surface. Check that the mica insulators, which lie between the segments of the commutator, are undercut. The standard groove depth is 0·8 – 1·0 mm (0·03 – 0·04 in) but if the average groove depth is less than this the armature should be renewed or returned to a Suzuki Agent for re-cutting.

7 Replace the brushes in their holders and check that they slide quite freely. Make sure the brushes are replaced in their original positions because they will have worn to the profile of the commutator. Replace and tighten the end cover, then replace the starter motor and cable.

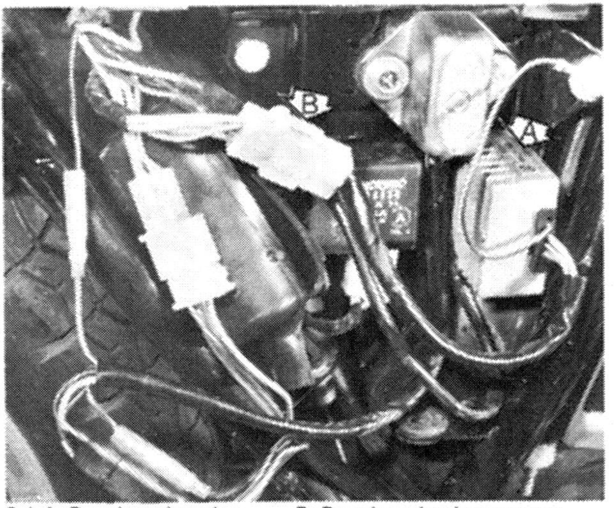

3.1 A: Regulator location B: Regulator lead connector

8.1 Fuse is located beneath dual seat (X7)

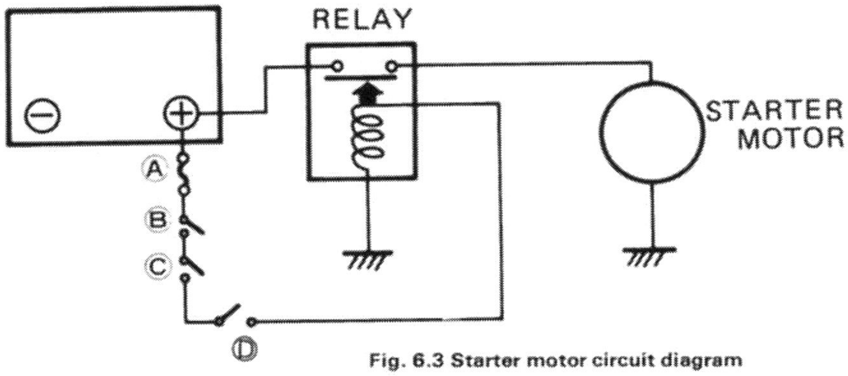

RELAY

STARTER MOTOR

A Main fuse
B Ignition switch
C Engine kill switch
D Starter switch

Fig. 6.3 Starter motor circuit diagram

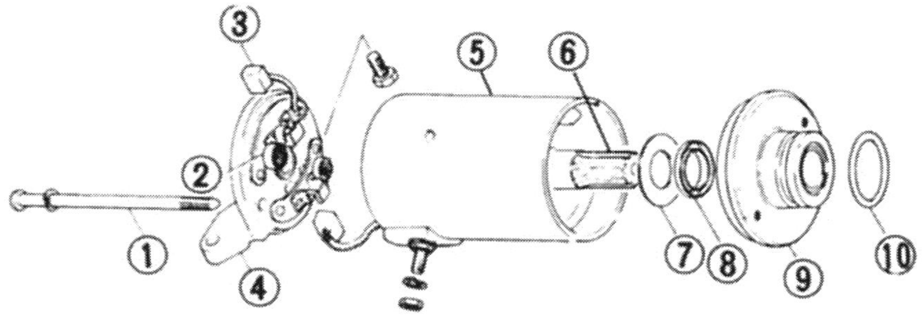

1 Screw – 2 off
2 Spring – 2 off
3 Brush – 2 off
4 End cover
5 Main body
6 Armature
7 Shim
8 Oil seal
9 End cover
10 O-ring

Fig. 6.4 Starter motor – component parts

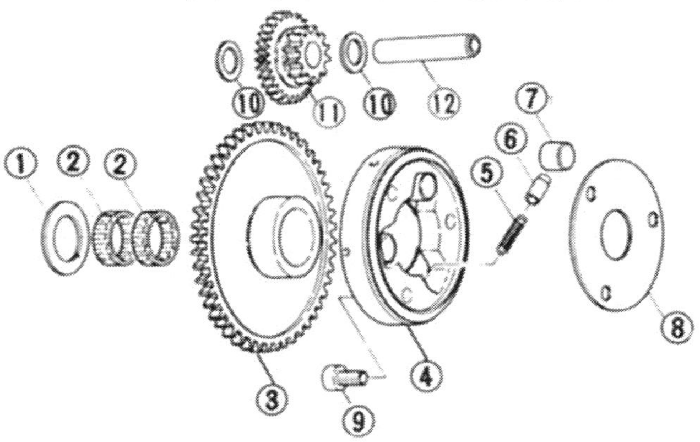

1 Thrust washer
2 Bearing – 2 off
3 Starter clutch gear
4 Clutch housing
5 Spring – 3 off
6 Plunger – 3 off
7 Roller – 3 off
8 Back plate
9 Screw – 3 off
10 Thrust washer
11 Idler pinion
12 Support shaft

Fig. 6.5 Starter clutch components

10 Starter solenoid switch: function and location

1 The starter motor switch is designed to work on the electro-magnetic principle. When the starter motor button is depressed, current from the battery passes through windings in the switch solenoid and generates an electro-magnetic force which casues a set of contact points to close. Immediately the points close, the starter motor is energised and a very heavy current is drawn from the battery.

2 This arrangement is used for at least two reasons. Firstly, the starter motor current is drawn only when the button is depressed and is cut off again when pressure on the button is released. This ensures minimum drainage on the battery. Secondly, if the battery is in a low state of charge, there will not be sufficient current to cause the solenoid contacts to close. In consequence, it is not possible to place an excessive drain on the battery which in some circumstances can cause the plates to overheat and shed their coatings. If the starter will not operate, first suspect a discharged battery. This can be checked by trying the horn or switching on the lights. If this check shows the battery to be in good shape, suspect the starter switch which should come into action with a pronounced click. It is located behind the right-hand side panel and can be identified by the heavy duty starter cable connected to it. It is not possible to effect a satisfactory repair if the switch malfunctions; it must be renewed.

11 Headlamp: replacing the bulbs and adjusting beam height

1 In order to gain access to the headlamps bulbs it is necessary to first remove the rim, complete with the reflector and headlamp glass. The rim is retained by two screws which pass through the headlamp shell just below the two headlamp mounting bolts.

2 The bulbholder assembly is secured to the headlamp reflector by a rubber cap, and this can be peeled away to facilitate removal of the bulb. To release the bulb, it should be pushed inwards slightly, then turned anti-clockwise. A reflector that accepts a pilot bulb is fitted to all models delivered to countries or states where parking lights are a statutory requirement. The pilot bulb is held in the bulb holder by a bayonet fixing.

3 Beam height on all models is effected by tilting the headlamp shell after the mounting bolts have been loosened slightly. In the UK, regulations stipulate that the headlamps must be arranged so that the light will not dazzle a person standing at a distance greater than 25 feet from the lamp, whose eye level is not less than 3 feet 6 inches above that plane. It is easy to approximate this setting by placing the machine 25 feet away from a wall, on a level road, and setting the beam height so that it is concentrated at the same height as the distance of the centre of the headlamp from the ground. The rider must be seated normally during this operation and also the pillion passenger, if one is carried regularly.

12 Stop and tail lamp: replacement of bulbs

1 The combined stop and tail lamp bulb contains two filaments, one for the stop lamp and one for tail lamp.

2 The offset pin bayonet fixing bulb can be renewed after the plastic lens cover and screws have been removed.

13 Flashing indicator lamps: replacing bulbs

1 Flashing indicator lamps are fitted to the front and rear of the machine. They are mounted on short stalks through which the wires pass. Access to each bulb is gained by removing the two screws holding the plastic lens cover. The bulbs are retained by a bayonet fixing.

11.1 Headlamp is retained by two screws

11.2a Turn bulbholder anticlockwise ...

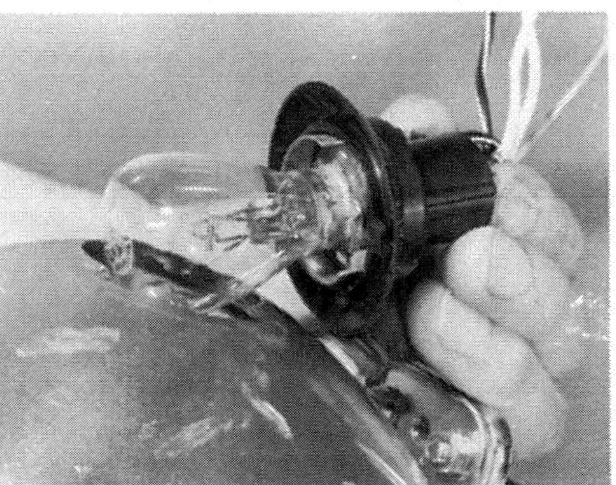

11.2b ... to release it from reflector

11.2c Parking lamp is a bayonet fitting

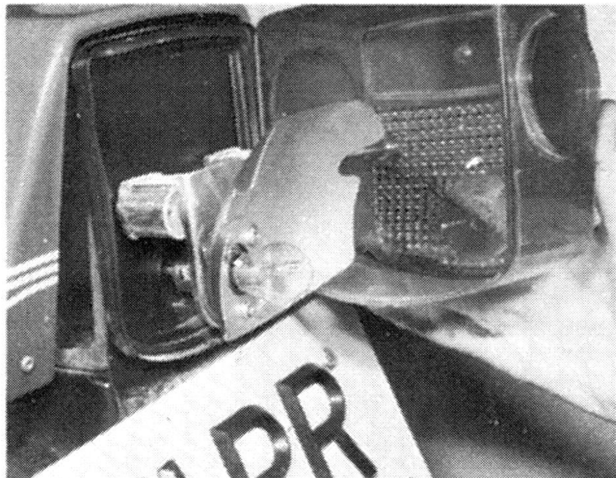

12.2 Remove lens to gain access to stop/tail bulb

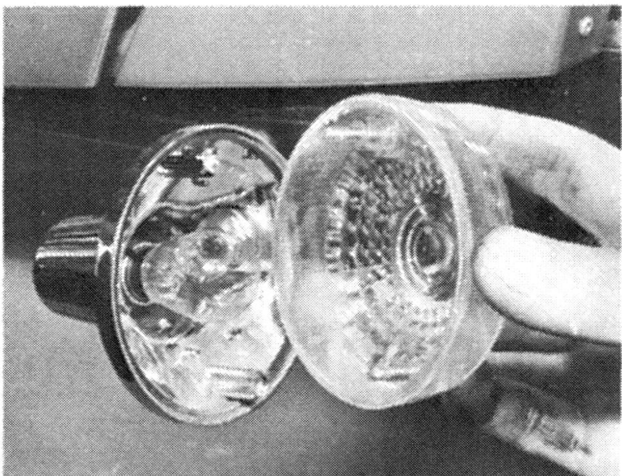
13.1 Indicator lenses are retained by two screws

14 Flashing indicator relay: location and replacement

1 The flashing indicator relay fitted in conjunction with the flashing indicator lamps is located behind the rectifier, behind the left-hand side cover. It is mounted in a rubber 'box' which isolates it from the harmful effects of vibration.
2 When the relay malfunctions, it must be renewed; a repair is impracticable. When the unit is in working order audible clicks will be heard which coincide with the flash of the indicator lamps. If the lamps malfunction, check firstly that a bulb has not blown, or the handlebar switch is not faulty. The usual symptom of a fault is one initial flash before the unit goes dead.
3 Take great care when handling a flasher unit. It is easily damaged, if dropped.

15 Warning and indicator lamps: bulb replacement

1 All bulbs fitted to the instrument heads or warning lamp console are of the bayonet fit type.
2 To gain access to the bulbs commence by removing the instrument console cover which is held at the periphery by screws. The central cluster of bulbs (where fitted) may be removed from their holders, one at a time. To allow removal of the instrument mounted bulbs, the drive cable on the instrument head must be detached and the two mounting nuts removed from the base of the casing. The instrument cover can then be lifted away, sufficiently to allow the bulb holder in question to be pulled out.

16 Switches and connectors: testing and repairs

1 In the event of any electrical fault, attention should be directed at the switch contacts and wiring connections in the circuit concerned. After long periods, these may become badly corroded, setting up a high resistance in the circuit and impairing the operation of the component controlled by them. If a multimeter is available, each switch connection and its associated wiring can be tested with the meter set on resistance.
2 The colour wiring diagrams at the end of this Chapter show the various switches in diagrammatic form, the lines denoting which contacts are connected in the various positions. With the machine's battery disconnected, check each of the relevant switch positions. When the two leads being checked are connected, the meter should indicate a completed circuit, with no discernible resistance.
3 The switch contacts can be kept clean and free of corrosion by using one of the proprietary switch cleaning fluids now available in aerosol form. These components can be injected into the switch body by means of the extension nozzle supplied, and require minimal dismantling. This form of preventative maintenance is especialy useful during winter, when the switches are most prone to corrosion.
4 In the event that a switch unit fails completely, little can be done to effect a satisfactory repair. The handlebar switches can be dismantled to a certain extent, but repair is limited to cleaning the switch contacts with fine abrasive paper, where access is possible. In many instances, renewal will be the only course of action.

17 Horn: location and examination

1 The horn is suspended from a flexible steel strip mounted on the cross member that joins the two front down tubes of the frame, immediately below the steering head. The flexible steel strip isolates the horn from the undesirable effects of high frequency vibration.
2 The horn has no external means of adjustment. If it malfunctions it must be replaced; it is a statutory requirement that the machine must be fitted with a horn that is in working order.

14.1 Indicator relay is mounted behind side panel

15.2a Remove drive cable and shroud ...

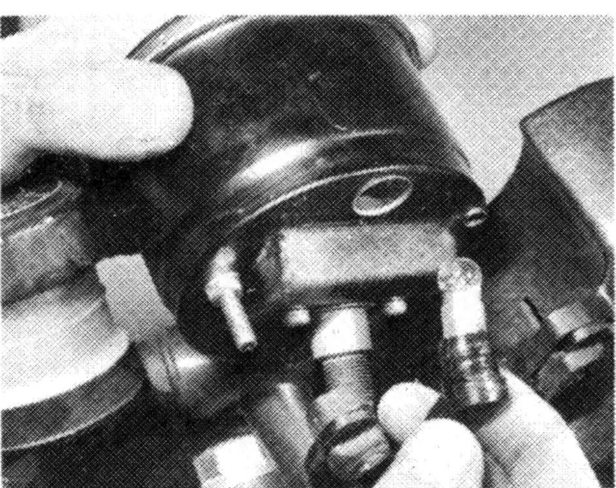
15.2b ... to gain access to instrument bulbs

16.4a Switch clusters clamp around handlebar ends ...

16.4b ... and may be separated for maintenance

18 Wiring: layout and examination

1 The wiring harness is colour-coded and will correspond with the accompanying diagrams. Where socket connectors are used they are designed so that reconnection can be made only in the one correct position.

2 Visual inspection will show whether any breaks or frayed outer coverings are giving rise to short circuits. Another source of trouble may be the snap connectors and sockets, where the connector has not been pushed home fully in the outer housing.

3 Intermittent short circuits can often be traced to a chafed wire that passes through or is close to a metal component, such as a frame member. Avoid tight bends in the wire or situations where the wire can become trapped between casings.

19 Ignition and lighting switch

1 The ignition and lighting switch is combined in one unit. It is operated by a key, which cannot be removed when the ignition is switched on.

2 The number stamped on the key will match also the number of the steering head lock and the filler cap lock. A replacement key can be obtained if the number is quoted; if either lock or the ignition switch is changed extra keys will be needed.

3 It is not practicable to repair the ignition switch if it malfunctions. It should be replaced with a new lock and key to suit.

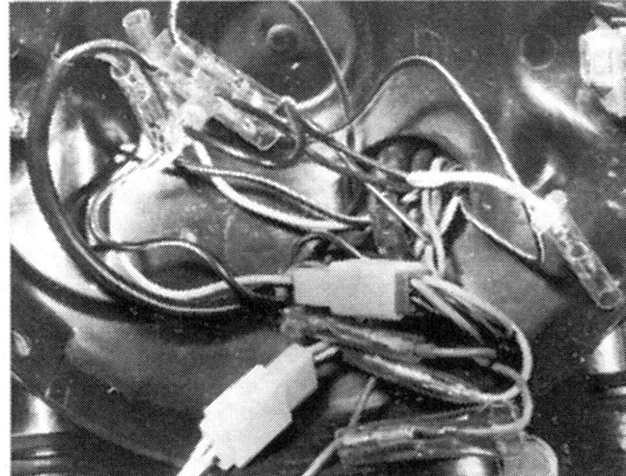
18.2a Wiring is colour-coded for easy identification

18.2b Connectors and terminals should be kept corrosion-free

19.1 Ignition switch is mounted in instrument panel (X7)

20 Fault diagnosis: electrical system

Symptom	Cause	Remedy
Complete electrical failure	Blown fuse	Check wiring and electrical components for short circuit before fitting new 15 amp fuse.
	Isolated battery	Check battery connections, also whether connections show signs of corrosion.
Dim lights, horn inoperative	Discharged battery	Recharge battery with battery charger and check whether alternator is giving correct output (electrical specialist)
Constantly 'blowing' bulbs	Vibration, poor earth connection	Check whether bulb holders are secured correctly. Check earth return or connections to frame.
	Defective regulator/rectifier	Test and renew as required

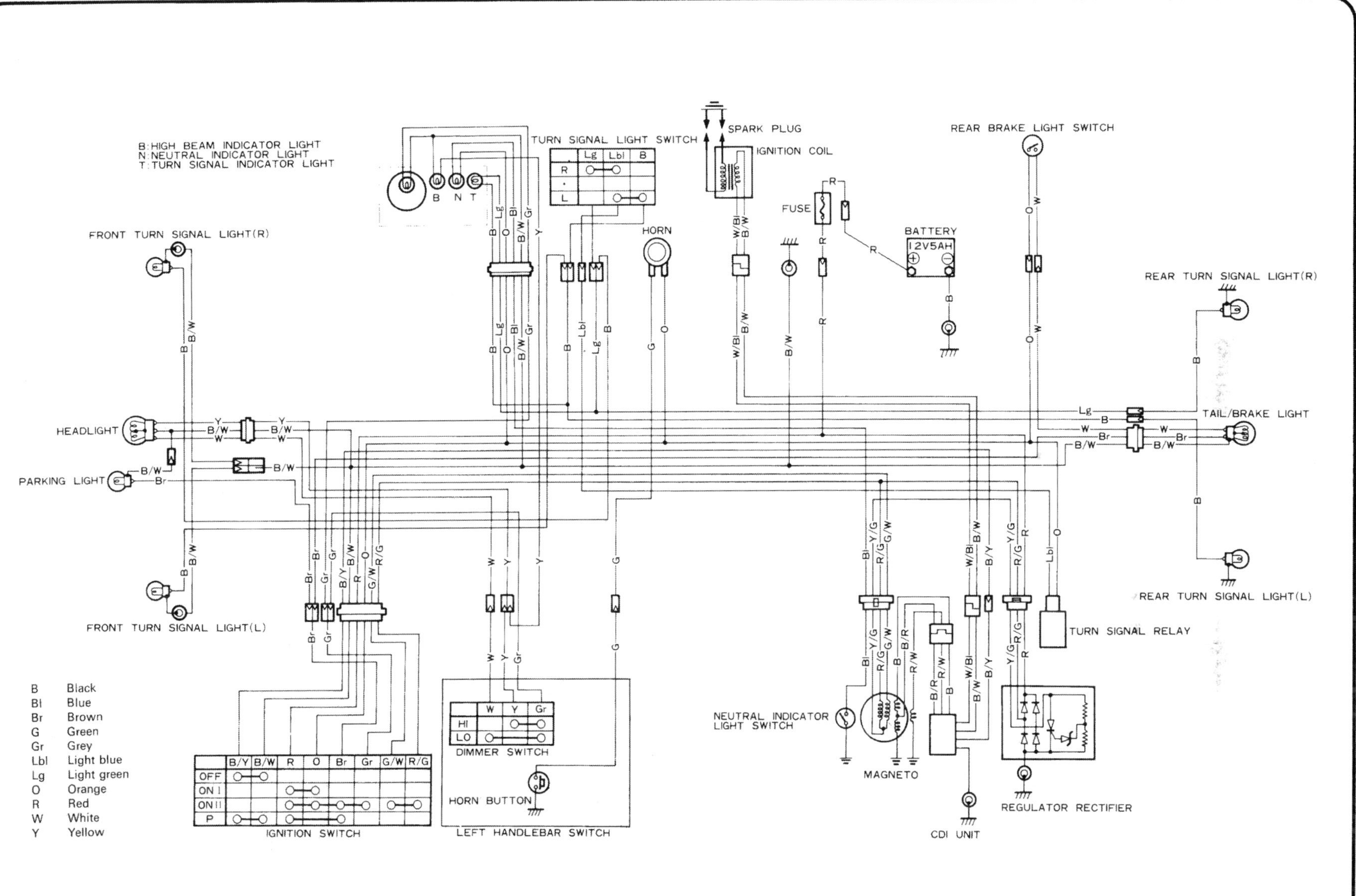

Wiring diagram – SB200

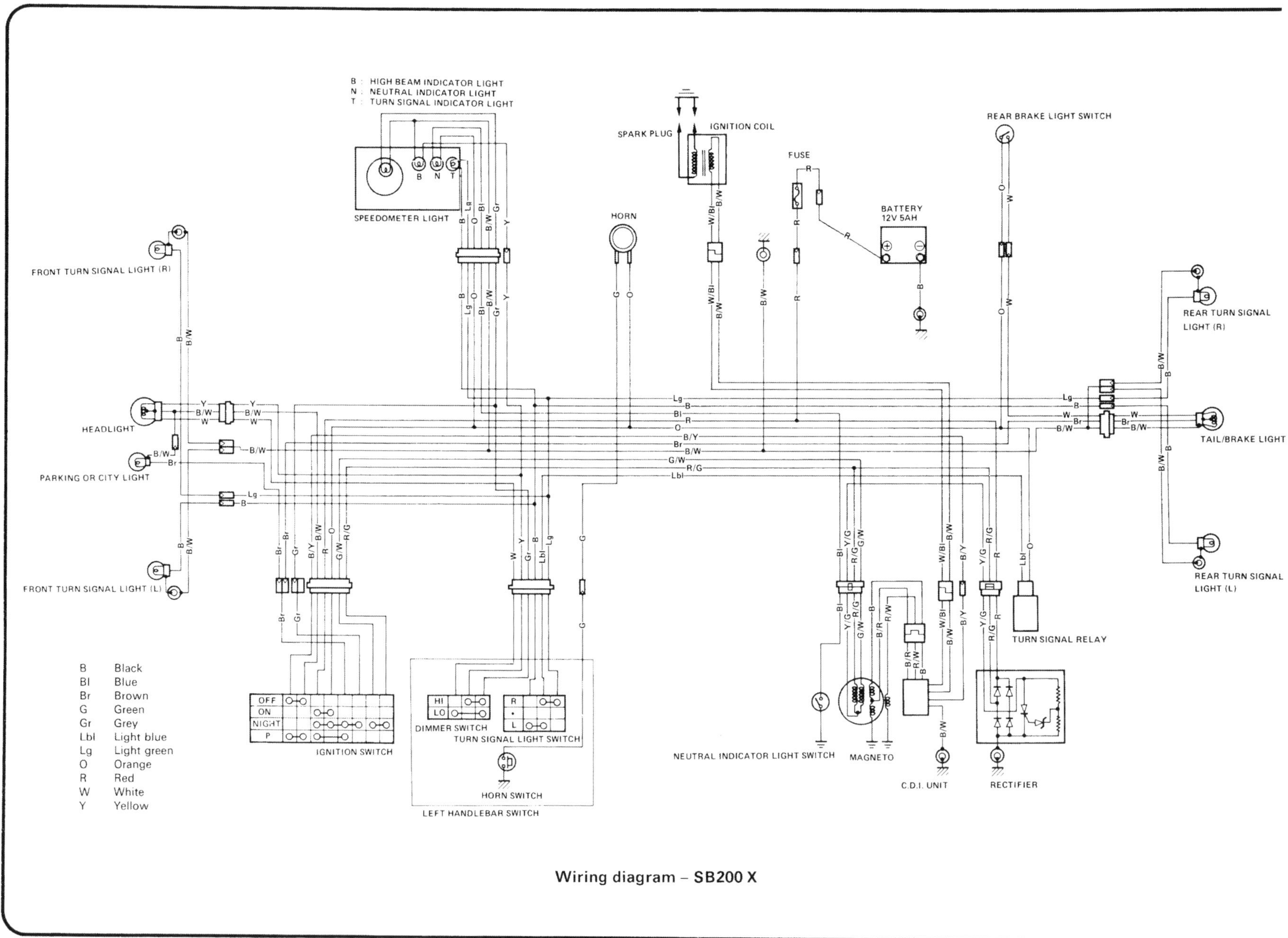

Wiring diagram – SB200 X

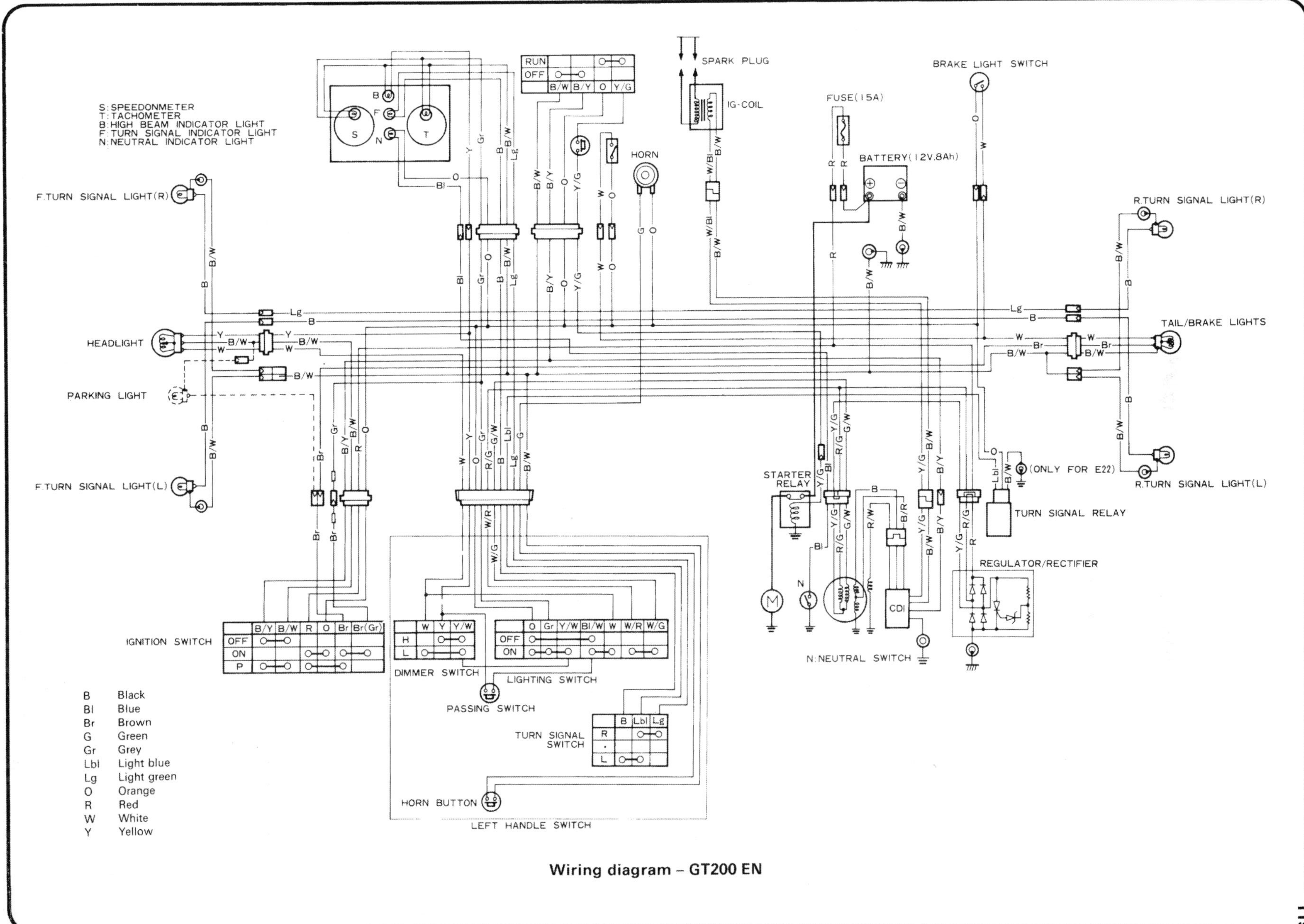

Wiring diagram – GT200 EN

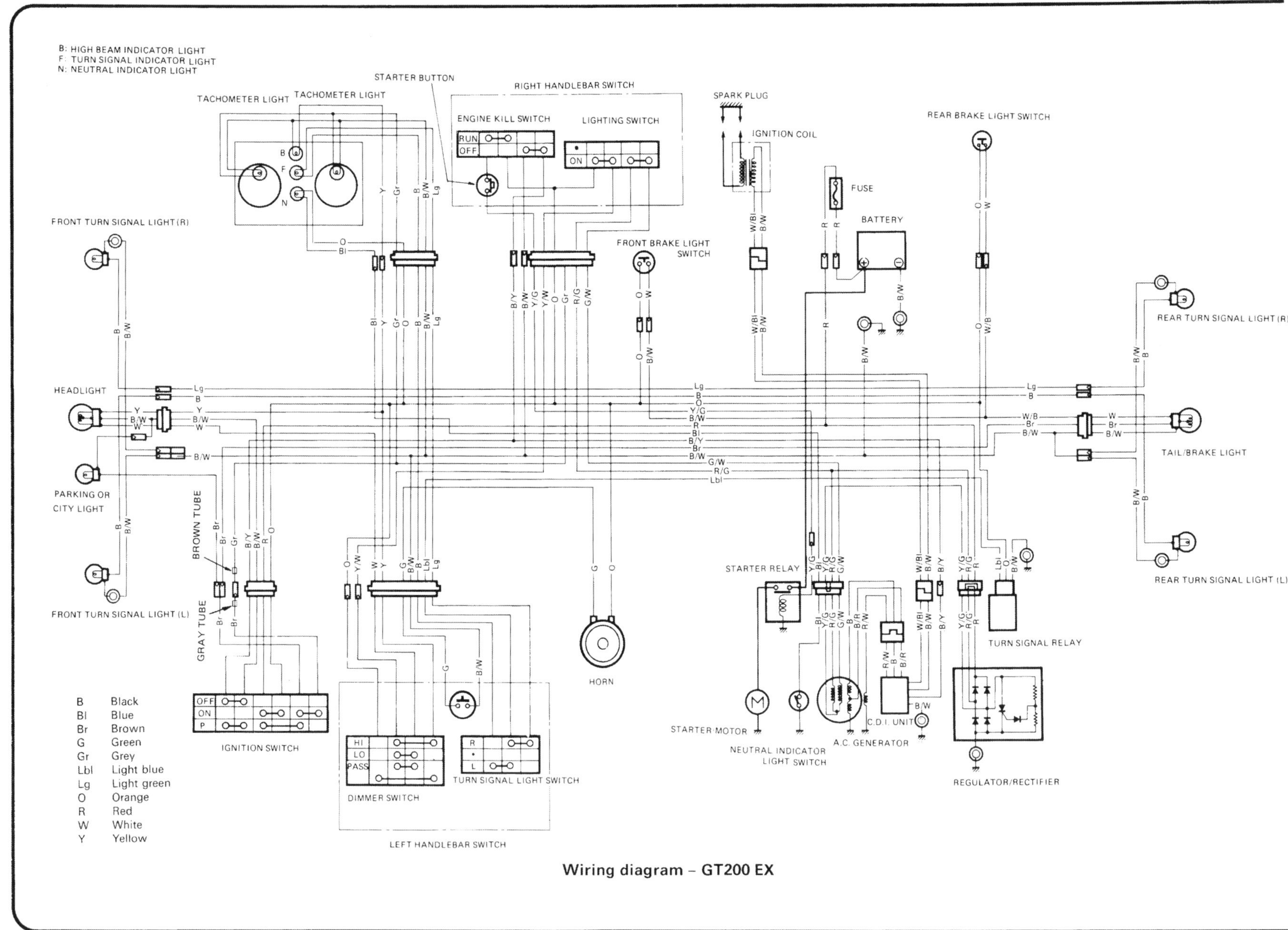

Wiring diagram – GT200 EX

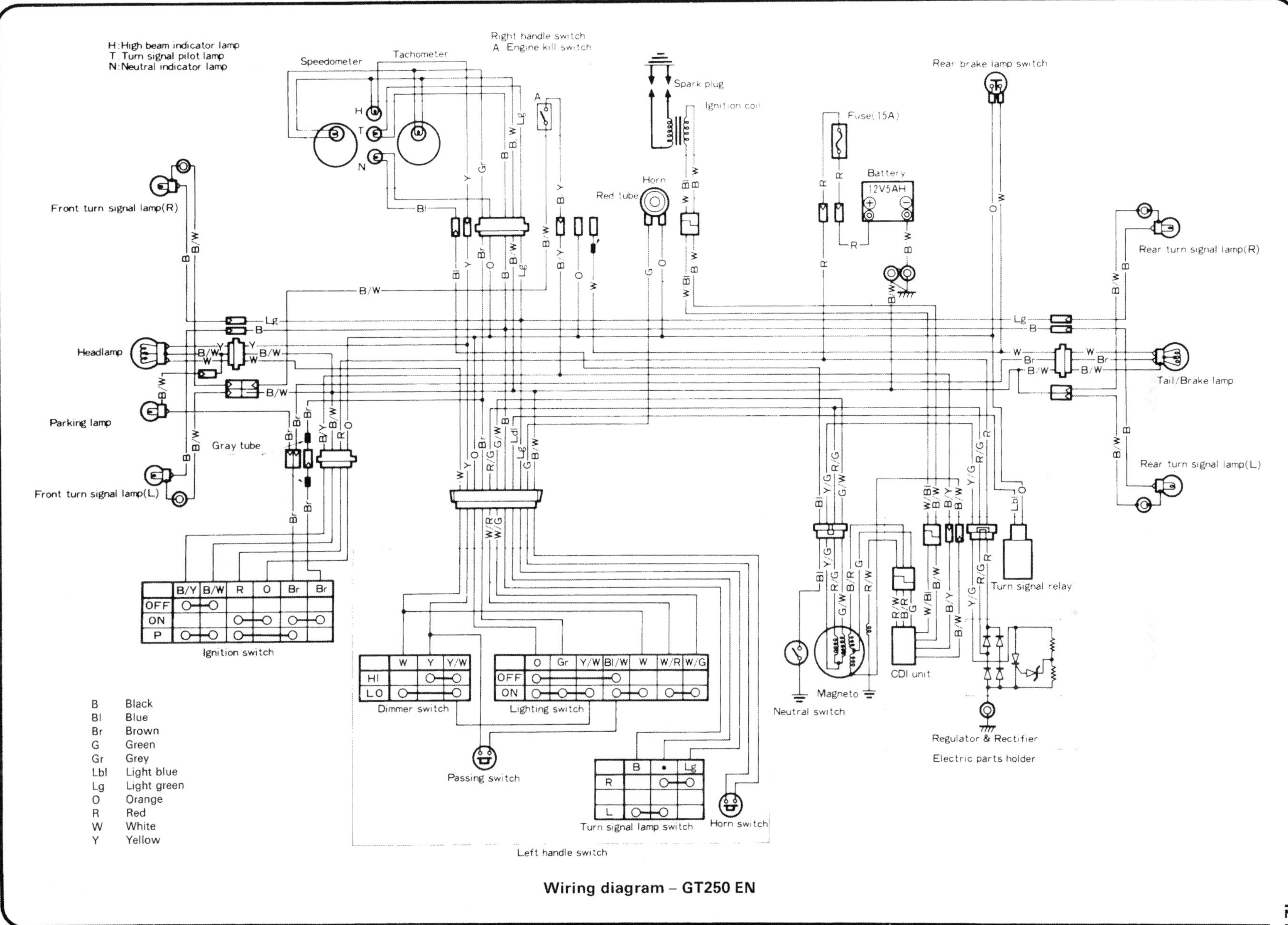

Wiring diagram – GT250 EN

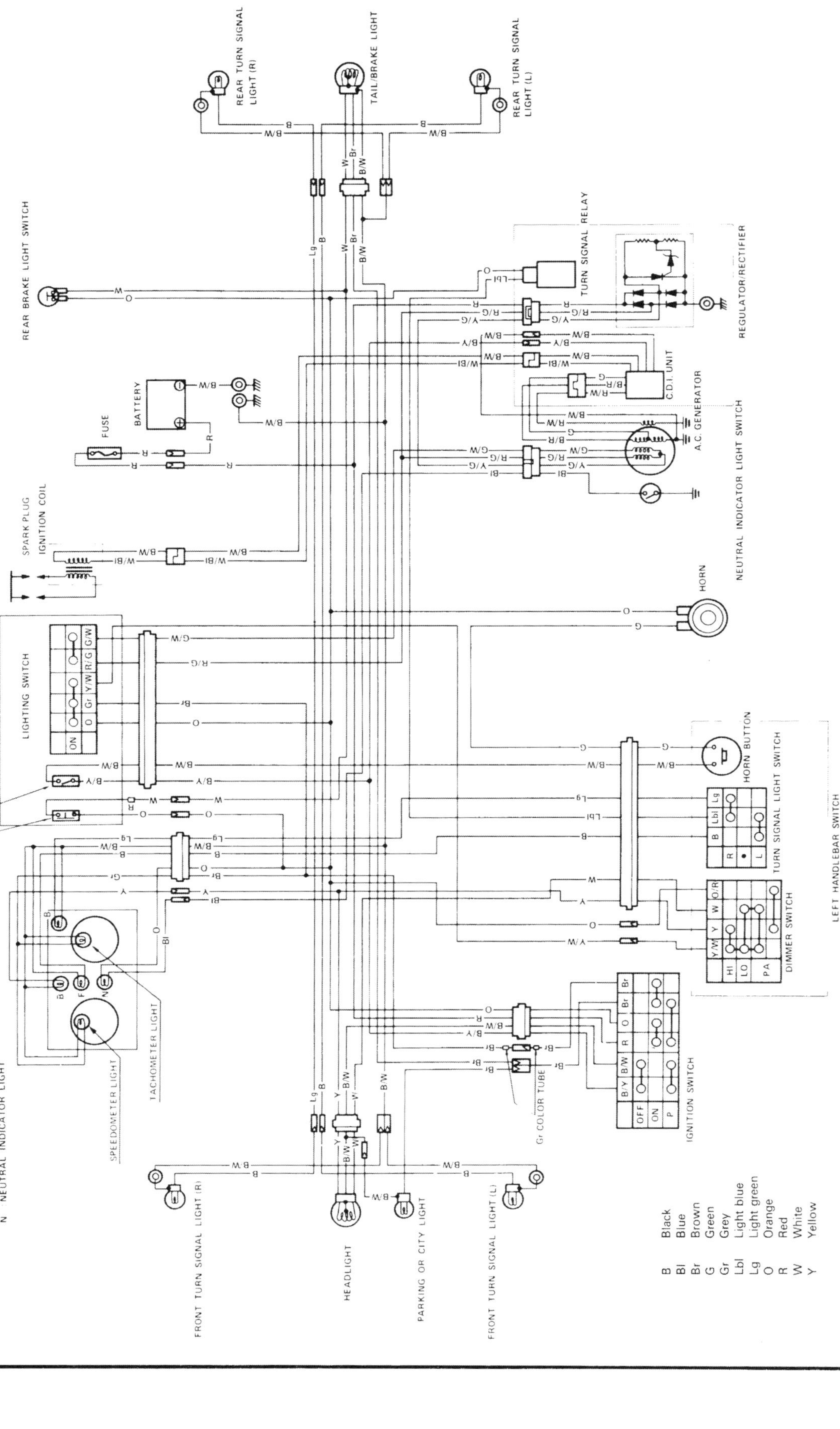

B HIGH BEAM INDICATOR LIGHT
F TURN SIGNAL INDICATOR LIGHT
N NEUTRAL INDICATOR LIGHT

FRONT BRAKE LIGHT SWITCH
ENGINE KILL SWITCH
RIGHT HANDLEBAR SWITCH
LIGHTING SWITCH

ON	O	Gr	Y/W	R/G	G/W

SPARK PLUG
IGNITION COIL
REAR BRAKE LIGHT SWITCH
FUSE
BATTERY
SPEEDOMETER LIGHT
TACHOMETER LIGHT
FRONT TURN SIGNAL LIGHT (R)
REAR TURN SIGNAL LIGHT (R)
HEADLIGHT
TAIL/BRAKE LIGHT
PARKING OR CITY LIGHT
FRONT TURN SIGNAL LIGHT (L)
REAR TURN SIGNAL LIGHT (L)
Gr COLOR TUBE
TURN SIGNAL RELAY

	B/Y	B/W	R	O	Br	Br
OFF						
ON						
P						

IGNITION SWITCH

	Y/W	Y	W	O/R
HI				
LO				
PA				

DIMMER SWITCH

	B	Lbl	Lg
R			
•			
L			

TURN SIGNAL LIGHT SWITCH
HORN BUTTON
LEFT HANDLEBAR SWITCH
HORN
NEUTRAL INDICATOR LIGHT SWITCH
A.C. GENERATOR
C.D.I. UNIT
REGULATOR/RECTIFIER

B Black
Bl Blue
Br Brown
G Green
Gr Grey
Lbl Light blue
Lg Light green
O Orange
R Red
W White
Y Yellow

Wiring diagram – GT250 EX

Metric conversion tables

			Millimetres to Inches		Inches to Millimetres	
Inches	**Decimals**	**Millimetres**	**mm**	**Inches**	**Inches**	**mm**
1/64	0.015625	0.3969	0.01	0.00039	0.001	0.0254
1/32	0.03125	0.7937	0.02	0.00079	0.002	0.0508
3/64	0.046875	1.1906	0.03	0.00118	0.003	0.0762
1/16	0.0625	1.5875	0.04	0.00157	0.004	0.1016
5/64	0.078125	1.9844	0.05	0.00197	0.005	0.1270
3/32	0.09375	2.3812	0.06	0.00236	0.006	0.1524
7/64	0.109375	2.7781	0.07	0.00276	0.007	0.1778
1/8	0.125	3.1750	0.08	0.00315	0.008	0.2032
9/64	0.140625	3.5719	0.09	0.00354	0.009	0.2286
5/32	0.15625	3.9687	0.1	0.00394	0.01	0.254
11/64	0.171875	4.3656	0.2	0.00787	0.02	0.508
3/16	0.1875	4.7625	0.3	0.01181	0.03	0.762
13/64	0.203125	5.1594	0.4	0.01575	0.04	1.016
7/32	0.21875	5.5562	0.5	0.01969	0.05	1.270
15/64	0.234375	5.9531	0.6	0.02362	0.06	1.524
1/4	0.25	6.3500	0.7	0.02756	0.07	1.778
17/64	0.265625	6.7469	0.8	0.03150	0.08	2.032
9/32	0.28125	7.1437	0.9	0.03543	0.09	2.286
19/64	0.296875	7.5406	1	0.03937	0.1	2.54
5/16	0.3125	7.9375	2	0.07874	0.2	5.08
21/64	0.328125	8.3344	3	0.11811	0.3	7.62
11/32	0.34375	8.7312	4	0.15748	0.4	10.16
23/64	0.359375	9.1281	5	0.19685	0.5	12.70
3/8	0.375	9.5250	6	0.23622	0.6	15.24
25/64	0.390625	9.9219	7	0.27559	0.7	17.78
13/32	0.40625	10.3187	8	0.31496	0.8	20.32
27/64	0.421875	10.7156	9	0.35433	0.9	22.86
7/16	0.4375	11.1125	10	0.39370	1	25.4
29/64	0.453125	11.5094	11	0.43307	2	50.8
15/32	0.46875	11.9062	12	0.47244	3	76.2
31/64	0.484375	12.3031	13	0.51181	4	101.6
1/2	0.5	12.7000	14	0.55118	5	127.0
33/64	0.515625	13.0969	15	0.59055	6	152.4
17/32	0.53125	13.4937	16	0.62992	7	177.8
35/64	0.546875	13.8906	17	0.66929	8	203.2
9/16	0.5625	14.2875	18	0.70866	9	228.6
37/64	0.578125	14.6844	19	0.74803	10	254.0
19/32	0.59375	15.0812	20	0.78740	11	279.4
39/64	0.609375	15.4781	21	0.82677	12	304.8
5/8	0.625	15.8750	22	0.86614	13	330.2
41/64	0.640625	16.2719	23	0.09551	14	355.6
21/32	0.65625	16.6687	24	0.94488	15	381.0
43/64	0.671875	17.0656	25	0.98425	16	406.4
11/16	0.6875	17.4625	26	1.02362	17	431.8
45/64	0.703125	17.8594	27	1.06299	18	457.2
23/32	0.71875	18.2562	28	1.10236	19	482.6
47/64	0.734375	18.6531	29	1.14173	20	508.0
3/4	0.75	19.0500	30	1.18110	21	533.4
49/64	0.765625	19.4469	31	1.22047	22	558.8
25/32	0.78125	19.8437	32	1.25984	23	584.2
51/64	0.796875	20.2406	33	1.29921	24	609.6
13/16	0.8125	20.6375	34	1.33858	25	635.0
53/64	0.828125	21.0344	35	1.37795	26	660.4
27/32	0.84375	21.4312	36	1.41732	27	685.8
55/64	0.859375	21.8281	37	1.4567	28	711.2
7/8	0.875	22.2250	38	1.4961	29	736.6
57/64	0.890625	22.6219	39	1.5354	30	762.0
29/32	0.90625	23.0187	40	1.5748	31	787.4
59/64	0.921875	23.4156	41	1.6142	32	812.8
15/16	0.9375	23.8125	42	1.6535	33	838.2
61/64	0.953125	24.2094	43	1.6929	34	863.6
31/32	0.96875	24.6062	44	1.7323	35	889.0
63/64	0.984375	25.0031	45	1.7717	36	914.4

Conversion factors

Length (distance)

Inches (in)	X	25.4	= Millimetres (mm)	X	0.0394	= Inches (in)
Feet (ft)	X	0.305	= Metres (m)	X	3.281	= Feet (ft)
Miles	X	1.609	= Kilometres (km)	X	0.621	= Miles

Volume (capacity)

Cubic inches (cu in; in^3)	X	16.387	= Cubic centimetres (cc; cm^3)	X	0.061	= Cubic inches (cu in; in^3)
Imperial pints (Imp pt)	X	0.568	= Litres (l)	X	1.76	= Imperial pints (Imp pt)
Imperial quarts (Imp qt)	X	1.137	= Litres (l)	X	0.88	= Imperial quarts (Imp qt)
Imperial quarts (Imp qt)	X	1.201	= US quarts (US qt)	X	0.833	= Imperial quarts (Imp qt)
US quarts (US qt)	X	0.946	= Litres (l)	X	1.057	= US quarts (US qt)
Imperial gallons (Imp gal)	X	4.546	= Litres (l)	X	0.22	= Imperial gallons (Imp gal)
Imperial gallons (Imp gal)	X	1.201	= US gallons (US gal)	X	0.833	= Imperial gallons (Imp gal)
US gallons (US gal)	X	3.785	= Litres (l)	X	0.264	= US gallons (US gal)

Mass (weight)

Ounces (oz)	X	28.35	= Grams (g)	X	0.035	= Ounces (oz)
Pounds (lb)	X	0.454	= Kilograms (kg)	X	2.205	= Pounds (lb)

Force

Ounces-force (ozf; oz)	X	0.278	= Newtons (N)	X	3.6	= Ounces-force (ozf; oz)
Pounds-force (lbf; lb)	X	4.448	= Newtons (N)	X	0.225	= Pounds-force (lbf; lb)
Newtons (N)	X	0.1	= Kilograms-force (kgf; kg)	X	9.81	= Newtons (N)

Pressure

Pounds-force per square inch (psi; lbf/in^2; lb/in^2)	X	0.070	= Kilograms-force per square centimetre (kgf/cm^2; kg/cm^2)	X	14.223	= Pounds-force per square inch (psi; lbf/in^2; lb/in^2)
Pounds-force per square inch (psi; lbf/in^2; lb/in^2)	X	0.068	= Atmospheres (atm)	X	14.696	= Pounds-force per square inch (psi; lbf/in^2; lb/in^2)
Pounds-force per square inch (psi; lbf/in^2; lb/in^2)	X	0.069	= Bars	X	14.5	= Pounds-force per square inch (psi; lbf/in^2; lb/in^2)
Pounds-force per square inch (psi; lbf/in^2; lb/in^2)	X	6.895	= Kilopascals (kPa)	X	0.145	= Pounds-force per square inch (psi; lbf/in^2; lb/in^2)
Kilopascals (kPa)	X	0.01	= Kilograms-force per square centimetre (kgf/cm^2; kg/cm^2)	X	98.1	= Kilopascals (kPa)
Millibar (mbar)	X	100	= Pascals (Pa)	X	0.01	= Millibar (mbar)
Millibar (mbar)	X	0.0145	= Pounds-force per square inch (psi; lbf/in^2; lb/in^2)	X	68.947	= Millibar (mbar)
Millibar (mbar)	X	0.75	= Millimetres of mercury (mmHg)	X	1.333	= Millibar (mbar)
Millibar (mbar)	X	0.401	= Inches of water (inH_2O)	X	2.491	= Millibar (mbar)
Millimetres of mercury (mmHg)	X	0.535	= Inches of water (inH_2O)	X	1.868	= Millimetres of mercury (mmHg)
Inches of water (inH_2O)	X	0.036	= Pounds-force per square inch (psi; lbf/in^2; lb/in^2)	X	27.68	= Inches of water (inH_2O)

Torque (moment of force)

Pounds-force inches (lbf in; lb in)	X	1.152	= Kilograms-force centimetre (kgf cm; kg cm)	X	0.868	= Pounds-force inches (lbf in; lb in)
Pounds-force inches (lbf in; lb in)	X	0.113	= Newton metres (Nm)	X	8.85	= Pounds-force inches (lbf in; lb in)
Pounds-force inches (lbf in; lb in)	X	0.083	= Pounds-force feet (lbf ft; lb ft)	X	12	= Pounds-force inches (lbf in; lb in)
Pounds-force feet (lbf ft; lb ft)	X	0.138	= Kilograms-force metres (kgf m; kg m)	X	7.233	= Pounds-force feet (lbf ft; lb ft)
Pounds-force feet (lbf ft; lb ft)	X	1.356	= Newton metres (Nm)	X	0.738	= Pounds-force feet (lbf ft; lb ft)
Newton metres (Nm)	X	0.102	= Kilograms-force metres (kgf m; kg m)	X	9.804	= Newton metres (Nm)

Power

Horsepower (hp)	X	745.7	= Watts (W)	X	0.0013	= Horsepower (hp)

Velocity (speed)

Miles per hour (miles/hr; mph)	X	1.609	= Kilometres per hour (km/hr; kph)	X	0.621	= Miles per hour (miles/hr; mph)

Fuel consumption

Miles per gallon, Imperial (mpg)	X	0.354	= Kilometres per litre (km/l)	X	2.825	= Miles per gallon, Imperial (mpg)
Miles per gallon, US (mpg)	X	0.425	= Kilometres per litre (km/l)	X	2.352	= Miles per gallon, US (mpg)

Temperature

Degrees Fahrenheit = (°C x 1.8) + 32

Degrees Celsius (Degrees Centigrade; °C) = (°F - 32) x 0.56

** It is common practice to convert from miles per gallon (mpg) to litres/100 kilometres (l/100km), where mpg (Imperial) x l/100 km = 282 and mpg (US) x l/100 km = 235*

Index

Zeitfracht Medien GmbH
Ferdinand-Jühlke-Straße 7
99095 Erfurt, Deutschland
produktsicherheit@kolibri360.de